CONTROLS AND AUTOMATION FOR FACILITIES MANAGERS

Efficient DDC Systems Implementation

CONTROLS AND AUTOMATION FOR FACILITIES MANAGERS

Efficient DDC Systems Implementation

VIKTOR BOED, CEM

CRC Press
Taylor & Francis Group
Boca Raton London New York

CRC Press is an imprint of the
Taylor & Francis Group, an **informa** business

Acknowledgments

I wish to express my appreciation to John Whitney for his expert reading and commentary on the manuscript. I am also indebted to Joe Chadwick for his helpful work on controls specifications.

Special thanks to Rosalie Cooke for manuscript editing, and to Roberto Meinrath for reviewing the book.

First published 1996 by Chilton Book Company

Published 2019 by CRC Press
Taylor & Francis Group
6000 Broken Sound Parkway NW, Suite 300
Boca Raton, FL 33487-2742

© 1996 by Viktor Boed
CRC Press is an imprint of Taylor & Francis Group, an Informa business

First issued in paperback 2019

No claim to original U.S. Government works

ISBN 13: 978-0-367-44844-8 (pbk)
ISBN 13: 978-0-8019-8722-9 (hbk)

Visit the Taylor & Francis Web site at
http://www.taylorandfrancis.com

and the CRC Press Web site at
http://www.crcpress.com

Designed by Jerry O'Brien

Library of Congress Cataloging-in-Publication Data

Boed, Viktor.
    Efficient DDC systems implementation / Viktor Boed.
        p.   cm.—(Controls and automation for facilities managers)
    Includes bibliographical references and index.
    ISBN 0-8019-8722-9
      1. Buildings—Environmental engineering—Quality control.
  2. Buildings—Automation. 3. Digital control systems. 4. Total
quality management. I. Title. II. Series.
TH6031.B64   1996
696—dc20   96-22644
                                                  CIP

To my wife, Vlasta, who had to put up with the many months of my withdrawal from regular life, and to our dog, Max, who was my constant (most of the time sleeping) companion during the writing of this book.

Contents

INTRODUCTION

Building habitats is as old as the human race. The industry, therefore, is traditional, but it is also sensitive to new requirements for occupancy comfort and safety, as well as to new industry trends in meeting the occupancy-related requirements.

The building occupants, when inconvenienced by being too cold, too hot, or by being exposed to poor indoor air quality, do not understand (as they need not) which component of the building or building systems contributes to the problem. They assume that the problem is a function of either the building design and/or the inability of the operation and maintenance (O&M) organizations. The way to improve the situation is to adopt a global, integrated approach to design, development, and O&M of building systems. Project managers should understand this approach, and be instrumental in assembling design teams which look far beyond the project completion, having in mind the continuous operation of the building. Such an approach alters the traditional approach to building design, construction, and maintenance, and the division between design and development on one side, and operation and maintenance on the other.

This book takes a new look at the existing situation, by adopting a "Global Commissioning/Total Quality Management (TQM) approach" to design, implementation, operation, and maintenance. This approach requires improved transfer of essential knowledge from the design and development teams to operations and maintenance personnel. This is the knowledge that guarantees continuous, trouble free, and economical operation of building systems. Simultaneously, with a team effort at every phase of the project, cumulative knowledge of all the participants can be funneled into the new design. A variety of participants is very important because the building environment is influenced by so many and such varied conditions, building components, and systems.

The complex web of design, engineering, development, construction, operation, and maintenance activities provided by many different organizations is aimed at a common goal: providing a safe and comfortable working environment for building occupants. Most of the building service activities are in one of three categories:

1. Design
2. Construction
3. Operations and Maintenance

Services for the first two categories are provided by outside organizations such as architectural and engineering organizations, construction management firms, and independent contractors. Building operation and maintenance is customarily done by the owner's in-house organizations such as facilities management.

Occupancy comfort, a primary requirement in this industry, is a function of the building and its environmental control systems. Among the major components of a modern building, environmental control systems are designed to provide and maintain safe and comfortable working conditions inside the building. Such conditions aid indoor activities and stimulate productivity. Environmental control systems are also required to control the building environment continuously, at its designed parameters, and at optimum operating cost.

The term *environmental control systems* refers to all associated Heating, Ventilation, and Air Conditioning systems (HVAC); building controls and automation systems; and other systems providing a safe and comfortable building environment. Building Automation Systems (BAS) such as Direct Digital Controls (DDC) systems are the "brains" controlling the connected building mechanical systems. Their application software contains definition of all connected field points and operating logic for controlled and monitored building systems. Note: BAS are synonymous with Energy Management and Control Systems (EMCS).

Another challenge the building industry is facing is its old, traditional approach to implementing Building Automation Systems. This approach has its roots in the single loop local pneumatic or electrical controllers of the past. Traditionally, controls were treated as any other trade on the job. They were installed, in many cases, by the job's electrical or mechanical contractors.

Introduction of computer-based control systems, especially remote intelligent DDCs, made this traditional approach less than suitable for the evolving building industry. New, complex building requirements demand the installation of more technologically-advanced systems. These systems control the building environment in much closer tolerances, and meet the need to conserve energy and reduce operating and maintenance costs.

The continuing evolution of technology in the building industry calls for alteration of past practices:

a. Provide the transfer of information from design and development to operations and maintenance. This can be accomplished by the participation of operations and maintenance personnel in the design and development process.

b. Integrate BAS design into the overall design process. This can be accomplished by fostering teamwork among HVAC engineers, BAS systems engineers, and facilities engineers throughout all phases of the project.

The main focus of this handbook is on DDC systems. They became one of the fastest growing components of the building industry. They have also become one of the best investments for the building owner, providing not only fast payback, but also improved building operations and maintenance. DDC systems are currently not utilized to their full potential. This book explores possibilities for their integration, and explains utilization of their features. Only operators who understand building characteristics and HVAC systems design, limitations, and operating logic, can use the DDC system to its full capacity.

This book looks at the entire process, having in mind the end result: quality building environment and occupancy comfort. It takes the owner's perspective in relation to the outside consulting and contracting world. This approach has been developed from two perspectives:

1. My own engineering and management experience which took me from consulting engineering, systems development, and product management of DDC systems for a major vendor, to facilities engineering for a major university;

2. My own experience with developing and implementing operating procedures for DDC systems. This changed my approach to project management and DDC systems implementation, operation, and maintenance.

The key components of these changes, which are called the Global Commissioning/Total Quality Management approach, are:

a. Creating site specifications
b. Establishing a partnership with DDC vendors
c. Reviewing significant stages of the design documentation and implementation
d. Applying input from O&M personnel to design, installation, and commissioning of the project.

These four components are the pillars of this book. They can be adopted for BAS implementations by most facilities. Implementation of the Global Commissioning/TQM approach does not increase significantly the project budget, yet the long-term benefits to the owner are enormous. There are no losers in the process. Architects and consulting engineers benefit from the team approach to design. Guidelines and a structured approach to DDC systems can shorten design time. Systems engineers can assist design teams with relevant knowledge which leads to even fuller utilization of DDC systems capabilities. BAS vendors and contractors benefit from the clear specifications, and overall participation in the design process. However, the greatest beneficiaries are the building operators, service technicians, and the occupants, who receive well-designed and operational environmental control systems.

CONTROLS AND AUTOMATION FOR FACILITIES MANAGERS

Efficient DDC Systems Implementation

Chapter 1

The Global Commissioning/Total Quality Management Approach

From Delivering the Building to Global Commissioning and Total Quality Management

Present Practices in Building Delivery

To understand present practices, you have to look at the building construction industry and the ways HVAC and DDC systems are designed and installed. Not that long ago, buildings were built with only a few manually-operated mechanical and electrical systems. The final turn-over and building delivery followed right after construction completion, cleaning up of the site, and testing of installed systems. But buildings have changed with additional requirements for safety, occupant comfort, and efficient operations, while reducing energy and operating costs. This is achieved with the addition of automated environmental control systems. These systems control the building environment in much tighter tolerances, meeting end requirements for safety and comfort. They must also meet demands for energy-efficient operation, and must contain the ever increasing cost for operation and maintenance. Modern buildings are like ships or airplanes. They are designed to function with, and be operated by, numerous systems that provide safety and occupancy comfort.

Each new building construction, major renovation, or HVAC and HVAC controls retrofit project is unique. In part, this is because man likes to build unique habitats. Designers apply their talents to novel public buildings, offices, laboratories, factory buildings, and other buildings designed for specific functions. This leads to a wide variation in how buildings look, what building systems are designed and implemented, and how systems work.

The building industry is very competitive because the owners want buildings delivered, and systems installed, for minimum cost and in the shortest possible time. The most widely-used approach to meet these requirements is competitive bidding. Bidding is required for public projects, and is widely accepted by the general building industry. The cost is easily justified because in the bidding process the lowest bidder who adheres to the provided specification is selected. The process encourages the bidders to provide systems for the lowest cost. Life cycle costing, systems reliability, maintainability, craftsmanship, and performance guarantees are often omitted from the evaluation process due to a lack of measures that would provide uniform evaluation and justification of submitted bids.

One of the problems with the bidding process as it relates to automation systems, is the omission of criteria for performance, operation, maintenance, system upgrades, and other criteria important for long-term systems operation and maintenance. For example, in the bidding process, DDC systems selections are based on comparing hardware and software specifications of individual systems and their costs. Degree of automation, quality of the application software, quality of systems engineering, systems performance guarantees, ease of generation and modification of application software, systems operating characteristics, upgradability of the system, networking, and other features so vital for DDC system operation and maintenance are often not evaluated during the bidding process. As a result, only obvious deficiencies related to performance and maintenance of the environmental control systems are detected during design, implementation, and turn-over phases of the project. Performance, operational, and maintenance deficiencies are often not detected until later, during regular systems operation, often after expiration of the warranty period.

Reasons for Poor Building Performance

Poor system performance results in complaints from the occupants. Most complaints from building occupants are related to the operation of environmental control systems, which are the HVAC and HVAC controls and automation systems. According to the published survey of the Architectural Institute of America, 90 percent of complaints are due to occupancy comfort, or rather, discomfort. Because the DDC system is the brain of environmental control systems, all those associated with systems design, implementation, operation, and maintenance should pay increased attention to these statistics.

New Ways to Achieve Quality Design and Implementation of Building Environmental Control Systems

To improve present practices, the industry has to analyze design, installation, and project management practices. Environmental control systems must meet design standards, and provide occupancy comfort and efficient operation under all seasonal and operating conditions. "Acceptance of building systems" refers to the process from initial design to turn-over of a functioning environmental control system in a building ready for full occupancy. To have a successful project, acceptance criteria should be defined and implemented throughout all phases of the design and implementation process. The method used to achieve this goal is called Global Commissioning/Total Quality Management (GC/TQM). The major difference between this method and the existing turn-over process is that GC/TQM tests criteria at every phase of the design and implementation process, not just at the end of installation.

Global Commissioning/Total Quality Management (GC/TQM) offers a solution which, if properly applied, enhances quality design and systems implementation, and provides a bridge between design and implementation on one side and operation and maintenance on the other side (see Figure 1-1.).

Analysis of building controls and automation jobs seems to indicate that there are three areas that repeatedly call for improvements:

1. Perception
2. Participation
3. Procedures

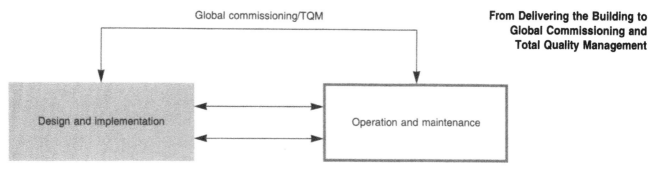

Figure 1-1. Global commissioning/total quality management

Perception

Building Automation Systems are *perceived* in the program definition and design phases to be part of the HVAC system's design. At these project phases, often there is no global systems definition to take advantage of distributed processing, networking, communications, management reporting, on-line maintenance problem analysis, and other native features of DDC controls and automation systems.

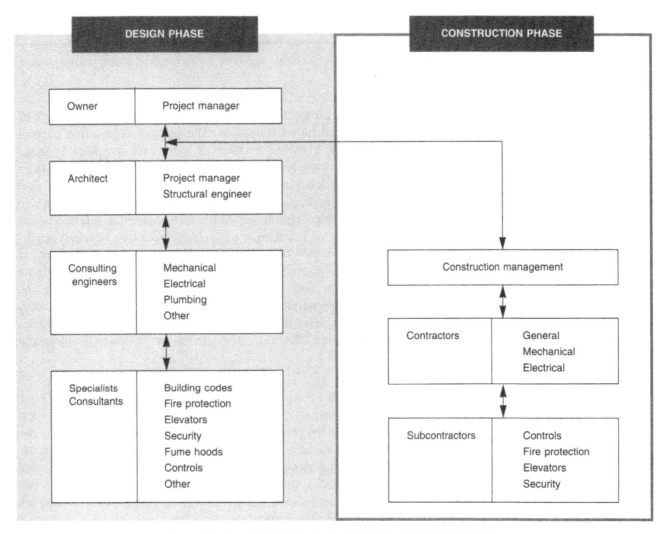

Figure 1-2. Project organization for new buildings and major renovations

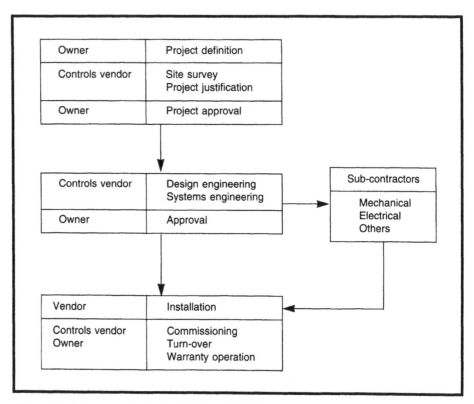

Figure 1-3. Project organization for HVAC and HVAC control retrofit jobs using controls vendors

Many engineers and project teams misrepresent the capabilities of the DDC system to the owner and future occupants. The general engineering population tends to (mentally) combine features of individual computer systems. Such perception leads to overrating off-the-shelf DDC systems' capabilities in overcoming possible deficiencies of the designed building and HVAC systems. Sometimes unrealistic, overly-ambitious presentations to the owner and decision-makers boost expectations before the DDC system is selected or designed for the application. The danger of such presentations becomes apparent as soon as the budget cuts take place. Often, the controls system is one of the first budget lines to be trimmed. The (incorrect) perception is that even a minimal control system will provide "all" expected features, laying the grounds for great disappointment in future systems performance. The only way to provide realistic expectations is to include the participation of experienced O&M personnel and DDC systems engineers in the design development process.

Participation

Participation of DDC systems engineers, and operations and maintenance personnel, is very limited at the programming and project design phases. Early design development is focused on architectural features, meeting the code requirements, and other big money items, rather than on environmental control systems, specifically DDC systems. Early design decisions at these phases that could impact facilities automation and DDC systems performance overall, are made without input from automation engineers, and without the participation of engineers with practical experience in systems operation and maintenance.

Why do you need such expertise? At the early stages of conceptualizing the building and its systems, it is easier to imagine the impact of a Greek marble lobby on the aesthetics of the building than to comprehend all ambient conditions, engineering calculations, and operating logic that goes into controlling the building at a constant room temperature of 70 degrees Fahrenheit year round. Review of design concepts from automation systems experts (engineers, operators, technicians) is necessary as feedback to architects' and HVAC engineers' ideas during these early stages of project definition. Engineers familiar with DDC systems design, operation and maintenance, should participate in the early design development phases of the project.

For many projects, DDC systems expertise does not become part of the review process until after bidding and final systems selection. The DDC systems "expert" is often the HVAC (consulting) engineer, who may or may not have expertise and operating experience with automation systems. The input of Operations and Maintenance is typically not solicited, evaluated, or noted until systems are installed, often during the final walkthrough, and prior to systems turnover for operation.

Design input originating from the practical experience of building operators, maintenance personnel, and systems engineers can be invaluable to the owner during the design development phases of the project. Many ills of environmental control systems operations are blamed on inadequate controls. These complaints can be reduced by incorporating systems engineering and operating expertise into the design process. Such expertise is based on experience gained in similar situations in the past. It is in the owner's and controls vendor's interest to enhance their participation in the design development phase of the project. After all, the system engineers and operations and maintenance personnel will stay with the installed system long after all other participants in the design development team have completed their job and moved on to other projects.

Procedures

Current design, development, and implementation practices lack procedures which would aid the owner and project manager in a step-by-step evaluation of design and implementation phases of the project.

One of the emerging approaches that could change the process is the Commissioning/Total Quality Management method. This approach, if properly developed for each phase of the project, influences the design and decision-making process. It keeps the end goal in the spotlight throughout all phases of design and the implementation process. Figures 1-2 and 1-3 depict two different project organizational charts. Figure 1-2 is for new building construction and major renovation. Figure 1-3 is for HVAC and HVAC control retrofit jobs using a full service controls vendor as a general contractor.

Main Characteristics of Global Commissioning/TQM Approach

The following discussions are focused on the Global Commissioning/Total Quality Management (GC/TQM) approach and its applicability to Building Automation Systems (BAS). TQM approaches and applications are well known, implemented, and documented for manufacturing and industry. The GC/TQM approach is new in its being applied to design and implementation of HVAC and DDC building systems. Whether the GC/TQM process is suitable for any individual project depends on the project itself, the structure of the design teams, and participants' familiarity with BAS design and implementation.

In general, the GC/TQM approach has three major building blocks:

1. Setting goals
2. Setting performance requirements and measures
3. Identifying external and internal customers

Setting Goals

- The overall goal of any project is to deliver buildings and systems which are built according to highest standards of quality and meet code requirements.

- The overall goal for environmental control systems is to continuously maintain environmental conditions at designed parameters for every controlled space, and do so at optimum operating, energy and maintenance cost.

- The overall goal for the Building Automation Systems (BAS) is to provide the highest level of automation for the building and its HVAC systems, to provide alarm and management reporting, and to interface with other automation systems installed at the owner's site(s).

Besides the overall goals, specific objectives should be set for individual phases of given projects. These should include both long-term performance-related issues, and short-term objectives aimed at milestones within project phases.

Setting Performance Requirements and Measures

Overall performance requirements and measures for environmental control systems are set by the design parameters for the entire building as well as individual controlled spaces.

To assure desired performance of the installed systems, performance requirements and measures should be established for every phase of the project. This should be similar to quality assurance in the manufacturing industry, where each product is checked for quality, achieved parameters, and tolerances at each phase of the production process, before moving the product to the next production phase. Only such rigorous quality measures can assure high quality and high performance products.

Applying Quality Assurance

Quality assurance can be adapted to environmental controls design and implementation by:

- Developing requirements and measures for every phase and significant benchmarks of the design and implementation process. For example, for a design phase, it can be related to submittals of specific pieces of HVAC documentation for review (calculations, hardware specification, I/O list, etc.), or requirements for design documentation review at its various stages of completion (50%, 75%, 100% completion), and incorporation of comments and suggestions into the following phase. Similar requirements can be set up for significant benchmarks. For example, review and approval of pre-ordered hardware, or inspection of installed VAV boxes and their controllers by the owner's O&M personnel prior to closing up the ceiling.

- Developing a review process which provides constructive suggestions and input, leading to enhanced design, enhanced quality, and lower cost of installation. The review process should provide ideas and positive feedback to the design engineers and contractors which, in turn, can enhance the quality of designed systems, engineering solutions, documentation, and systems implementation.

- Developing procedures that enhance the learning process. The greatest asset of every engineer is his or her experience resulting from comparing the original design intentions with the actual systems performance. This could be provided by the ongoing review process, and final evaluation after project completion. For the design team, this could translate to comparing the original environmental systems design to the performance of installed systems, reliability of operation and systems maintainability. Evaluating and implementing design and installation approaches that worked in the past into new projects is prudent for all involved in the process.

Identifying Internal and External Customers

One of the more challenging aspects of the Global Commissioning/ TQM concept is to identify internal and external customers for every major task, within and outside of the design and implementation teams. Identifying customers enhances inter-relationships among the team members. Professionals strive to deliver products and services of the highest quality to their customers.

For example, in sports, everyone understands the team concept. Each player on the team may receive the ball from any of the players. The player who has the ball may pass it to his or her teammates, score, or lose it to the opposing team. The quality of the game depends on the quality of the individual players and on the quality of the hand-off from one player to another. Players on the same team are "internal customers" to each other, with the overall objective being the quality of the game and the final score.

Formal, especially external customer relationships, are well defined by contractual obligations. Informal and internal customer relations (e.g., within the design team, or between the design and review teams), are much more difficult to define because the team members represent such a variety of knowledge and skills. For example, when a controls contractor is a sub-contractor to the mechanical contractor, a formal relationship has been established by the contract. However, the same controls contractor can have an informal customer relationship with a (third party) systems integrator who has been invited to participate in the design development of overall facilities automation and networking. No contract has been written that requires the controls contractor and the systems integrator to have a formal, contractual relationship. However, an informal relationship should be established, since each requires information or skills from the other.

In the same way, the project manager may ask a DDC systems engineer who is an employee of the controls vendor and is familiar with the owner's facili-

7

ties, to review the HVAC design documentation from a systems engineering point of view, even prior to awarding of the contract. An informal customer relationship has now been established among the project manager, the systems engineer, the controls vendor, and the owner's facilities representatives.

The following examples of informal customer relations illustrate the concept of "customers" in the GC/TQM approach:

- According to the site specification, the maintenance department has to approve the locations of field sensors and actuators for maintainability and accessibility. For this task, the maintenance department becomes a customer to the controls vendor installing the DDC hardware.

- During software generation, the owner's operations department has to review and approve the colorgraphic screens to assure uniformity with the existing colorgraphics on site. The operators become customers to the controls vendor for review and approval of operator interfaces.

- The facilities engineer is responsible for the degree of automation, alarm reporting, systems long-term development, and interfaces, thus becoming an informal customer to the controls vendor.

Once you identify who the internal and external customers are, try to extend the contractual relationships, especially those of external customers, into more informal relationships. Relationships between individual participants in the design development process are transitory and often limited to the duration of the project. Identifying customers for major activities within defined phases of the project is much more complex than, for example, in the manufacturing environment where defined customer relationships are stable for the duration of a manufacturing process that might last many years.

In the teamwork approach, there is a customer for every task and activity. Formally or informally, external and internal customers provide reviews and quality assurance for the design documentation and installed hardware and software at every phase of the process. However, it is important to note that the ultimate responsibility for systems design and implementation remains within the contractual obligations of the participating companies. The purpose of the customer reviews is mainly to provide constructive feedback and QA checks throughout the design and implementation phases.

This approach of identifying customers is parallel to the successful management practices published by Tom Peters and other management consultants. "Provide Top Quality, as Perceived by the Customer" is a chapter in Mr. Peters' book, *Thriving on Chaos*. In another book, *In Search of Excellence*, Tom Peters writes "...IBM measures internal and external customer satisfaction on a monthly basis. These measures account for a large share of incentive compensation, especially for senior management." What a great idea! Wouldn't it be nice to implement similar practices for projects related to buildings and environmental control systems?

Customer relationships should be identified for engineering, and also for project management activities between design development teams and the owner's O&M departments. Identifying internal and external customers for individual project phases is an important task in the Global Commissioning/TQM approach.

The Commissioning Agent

Role of the Commissioning Agent

To manage the Global Commissioning/TQM process, the owner should appoint a commissioning agent. He or she should be responsible for definition and coordination of the process throughout the design and implementation phases of the project. The commissioning agent should be part of the design team, and a partner to the project manager. With the commissioning agent on the team, the project manager's main focus is then on establishing and managing project teams and project schedules, coordinating activities, managing the budget, and administering the project overall. The commissioning agent's main focus is on quality. He or she aids in project management by defining goals, setting up quality measures, and establishing informal customer relations and teamwork.

The commissioning agent should be experienced in managing engineers of different backgrounds and should be well regarded among the participants for his or her management as well as engineering qualities. To be able to focus the project teams on quality, systems performance, and reduced operating and maintenance costs, the commissioning agent must be familiar with systems performance, operations and maintenance needs, and practices of the owner. The strong management and engineering qualities are very important because the design development process is not very disciplined, and many important decisions are made by committees and in meetings. In many instances, final decisions are a result of compromises, or are based on the technical understanding of the subject by participants with different levels of engineering qualifications and experience. Such decisions may satisfy the minimalist approach, but can harm the quality of the project.

The commissioning agent should have a good working knowledge of environmental control systems design and operation. This includes an understanding of characteristics, features, and limitations of DDC systems available on the market. Consequently, the commissioning agent must be able to convey such information to the design team, owner, and future occupants. This assures more realistic expectations, as well as full utilization of DDC systems capabilities in conjunction with the designated HVAC systems.

Tasks of the Commissioning Agent

In general terms, the commissioning agent has to:

1. Establish goals for the process
2. Establish performance measures
3. Identify customers
4. Manage the quality review process.

He or she has to maintain both a team effort and a win-win attitude throughout the project. If necessary, he or she has to identify in-house and outside experts, and provide additional resources to the original project team.

The commissioning agent's tasks are presented at the end of every chapter, to aid the commissioning agent in planning and executing the related activities in each phase. The following tasks represent some of the more significant duties at individual phases of design and implementation:

Design Phase

During the design phase, the commissioning agent should focus on quality design. Using the review process, and implementation of the review comments, the commissioning agent assures meeting design parameters, and also meeting performance and maintenance requirements. Adherence to site specifications, conventions and criteria is important for site standardization. Equally important are interfaces to other systems or adjacent spaces, and interoperability requirements for automation systems. Some of the requirements may not be identified in the project scope, for example, if it is limited to a partial renovation, but are important for continued facilities development. Additional related work and expenses outside of the original scope of work must be identified by the commissioning agent during the early stages of the planning process. Overall requirements, related to interfaces and interoperability in automation systems approved by the facilities and project management, should be included in the project documentation and be part of the individual contracts.

Installation Phase

During the installation phase of the project, the commissioning agent should provide periodical reviews, and assure participation of operation and maintenance personnel in the process. They should play a major role during system start-up, walk through, and the turn-over process. The commissioning agent's responsibilities should continue beyond the final phase of the construction process, into outlining adequate training for systems operators and maintenance personnel. He or she should also review the quality of as-built documentation and its turn-over to the O&M department.

After Construction Completion

Shortly after construction completion, the project manager and commissioning agent should schedule a meeting with the design and installation teams to document their experiences and compare the original design with the installed systems and their actual operation. Constructive feedback can be invaluable for future projects, and can enhance the team's experience. This is a unique opportunity to capture the experience of the professionals assembled for the project.

Beginning of Warranty

Perhaps the most important obligation of the commissioning agent after the beginning of the warranty period should be periodical review of the systems operation, performance, and maintenance requirements. Follow-up reviews should be scheduled with the building occupants as well as with the operation and maintenance personnel on a regular basis. Implementation of their suggestions during the warranty period will result in improved and more efficient environmental systems operation.

Before End of Warranty

Before the end of the warranty period, the commissioning agent should again follow up and document operating and maintenance experience with the installed systems, and get feedback from the occupants to evaluate the provided occupancy comfort. This process is aimed at enhancement of future design and implementation practices.

Selection of Commissioning Agents for Small Projects

For smaller projects or smaller facilities, appointment of a dedicated commissioning agent might impose a financial burden on the project budget. In that case, the role of commissioning agent can be carried out by the project manager. Another option is to find qualified people among the O&M staff. Every site has a qualified mechanic or a controls technician with a passion for his or her area of expertise. Such staff members are willing to share their experience and provide input to the design team when invited to do so.

Selection of Commissioning Agents for Large Projects

For larger sites or more involved environmental control systems projects, the commissioning agent should be an engineer with HVAC systems background and practical experience, and should have a good working knowledge of controls and automation systems applications. For larger sites and more complex installations, the commissioning agent also can be an independent, outside engineer. Many large facilities have developed long-term relationships with consulting engineers or building automation systems engineers who, over time, have developed expertise related to implemented environmental control systems and that owner's O&M practices and requirements. For example, the pharmaceutical industry requires compliance with "Current Good Manufacturing Practices (CGMP)." This means an engineer with specialized knowledge is necessary to keep the site in compliance with CGMP. Independent experts are good candidates for the function of commissioning agent.

One other option for large facilities is to develop in-house expertise, which is most beneficial for the owner if it can be used in a succession of projects. The commissioning agent's responsibilities can be assumed by in-house facilities engineer(s) trained in the quality management process. They are already responsible for facilities' and systems' long range goals, systems engineering, and engineering support of O&M departments. Many of their duties coincide with those of the commissioning agent.

The advantage of using in-house expertise is in time allocation. Quality assurance responsibilities related to most projects is not a full-time job. Assignment of commissioning agent's duties to in-house facilities engineers, especially during times of slow construction activities, develops in-house quality assurance expertise, and also assures continuity in systems design and implementation over time. This is especially important for sites with project activities provided by several engineering organizations and contractors over time. In response to safety, occupant comfort, energy efficiency, and efficient building operation requirements, environmental controls systems are getting more and more complex. Automation of non-HVAC facility functions is more and more common in modern buildings. Facilities engineers should be overseeing the development process in order to assure adherence to quality performance and long-range plans for their facilities.

Chapter 2

Site Standardization and DDC Specifications

Site Standardization

Site standardization is an effort toward the long-term, uniform development of design approaches, systems, materials, and operation and maintenance practices for a facility. Facilities, especially larger, multi-building facilities, should implement standards for design, systems, purchased material, and also for training, operation, and maintenance. Site standardization assures the continuous development of facilities systems, and their uniformity, which in turn translates to more productive maintenance, and reduced training, operating, and material costs. Many facilities have adopted standardization as a measure to contain their operating and maintenance costs.

Facilities undergo periodical renovation as they mature. Multi-building facilities experience ongoing construction activities as part of their renovation program, or because of a change of production, change of utilization, or change of the needs of their tenants or occupants. Every renovation reflects design philosophies, prevailing technology, preferences of design engineers, and the procurement process itself. Since every renovation and new construction activity is carried on as an independent undertaking, without adequate standardization the facility may end up with a variety of markedly different systems.

Standardization should begin with the development of a collection of site standards that provide guidelines for design engineers in their design approaches, selection of material for HVAC systems, and selection of DDC systems. DDC site performance and material specifications should be part of the collection of facilities standards.

Definition of long-term automation goals and standardization on controls and automation systems allows the owners to migrate their existing systems into newly implemented controls and automation systems and networks.

Site standardization of environmental control systems results in familiarity of the O&M personnel with the operating characteristics and related maintenance of the installed systems. Consequently, this translates into improved productivity, reduced training and inventory costs, and improved services for the building occupants.

Over time, many existing facilities have implemented controls and automation systems of various levels of sophistication, produced by different manufacturers. Whether these systems control pieces of equipment or larger HVAC systems, they were designed when sufficient maintenance and energy resources were dedicated to building operations. Eventually, buildings became more complex due to new energy, safety, and environmental requirements and regulations. At the same time there is pressure on facilities management to contain the cost of building operations and maintenance. To contain these costs, facilities have to automate their systems and operations.

Automation of building systems means implementing and integrating real time controls and automation systems. It also means integrating existing systems into the new automation networks. To be successful in this endeavor, facilities have to develop long-range plans and standardize systems, control strategies, hardware, and communications. Individual projects then have to follow these standards and specifications for their systems to be compatible with the rest of the facility's automation.

Containing costs for facilities operation means reducing the overhead by:

1. Standardization of systems, which will result in reduction of material (spare parts) cost, reduction of training costs, and reduction of labor costs due to familiarity of the operating and maintenance personnel with the implemented system and trouble-shooting methods.

2. Implementing technology which will not only control the building systems, but also provide automation and on-line diagnostics. This can be achieved only by a global approach to facilities automation, rather than designing and bidding individual systems for jobs. Standardization of technology, communications, and BAS systems features provides the avenue to increased productivity of operations and maintenance.

3. Focus on quality starting at the early design stage of any individual project. The design focus should be on systems operation and maintenance. To achieve the desired quality, facilities should have standards for design, systems implementation, and for ongoing quality reviews of the project.

4. Selecting systems based on Life Cycle Cost analysis rather than on a first cost basis.

Site standardization provides benefits to the owner as well as to systems designers and control vendors. Well-defined site standards, design criteria, and site specifications reduce the number of unknowns in the project, thus increasing the productivity of participating design engineers and controls vendors. Most importantly, it provides the owner with uniform systems, documentation, and a platform on which the facilities management can standardize its operation.

DDC Performance Specification

A Practical Guide to Preparing Controls and Automation Specifications

The DDC systems specification may be prepared by the owner, a consulting (HVAC) engineer, or a controls vendor. There are commonalities as well as unique issues in every specification regardless of the author.

The owner has a vested interest in specifying DDC systems in compliance with long-term systems development goals for meeting the minimum life cycle cost. Owners may charge consulting engineers to develop a DDC specification for a particular job, or to include the owner's specification as part of the larger body of construction documents. (Inclusion of the owner's DDC specifications into the construction documentation, or references to it in the HVAC section, may yield higher confidence and acceptance by the facilities management, design team, and construction management.)

The following DDC specifications are written from the owner's perspective. It is assumed that one of the owner's organizations, typically, the facilities engineering department is responsible for long-term development and implementation of facilities automation, including Building Automation Systems (BAS).

An "owner's" specification is included in this book to encourage owners to develop their own site specifications. This could lead to long-term BAS systems development, standardization, and the integration of existing and future facilities automation systems.

Another compelling reason to develop a site-specific specification is to focus on issues unique to the site. Unlike "generic" specifications, which may not describe fully the site-specific conditions and requirements, owner's specifications are tailored for the unique site conditions. This approach also gives us an opportunity to include references to external organizations, such as controls vendors, contractors, and consulting engineering organizations involved in HVAC design.

The following specifications (or their sections) can be easily modified and used for other real time systems, such as power plant controls and automation systems, utilities metering systems, and other microprocessor based systems utilized by the facilities.

Every specification is a "live" document which should be improved with new experience, approaches, technology, material, etc. Specifications should be revised and re-issued periodically to keep up with technology, approaches, site conditions, and updates.

Site specifications can be, and should be, written for large or small facilities. The effort put into writing such documents pays for itself many times over. DDC systems based on site specifications provide the desired automation in compliance with long-term site automation goals, and operating and maintenance conditions.

Note: The project manager and the commissioning agent should ensure that consulting engineering organizations involved in HVAC design will follow the owner's instructions even if they are accustomed to a different format of specifications.

The following text of a sample DDC Performance Specification is presented in a sans serif typeface (like this sentence). Each spec section is delineated by rules and a gray band in the outside margin. Narratives, notes, and explanations pertaining to individual sections are in a serif typeface (like this sentence), and follow the individual sections of the sample specification.

Organization of the Performance Specification

The organization of the specification is important to the parties responsible for contributing to, reviewing, and responding to the document. Each item or reference should appear in a particular section of the specification to be easily found. For example, all discussions related to warranties should appear in the warranty section, and the other sections include only references to the warranty section.

The Performance Specification is divided into three basic parts:

1. Part I includes general information.

2. Part II discusses systems and their performance requirements.

3. Part III explains how to execute the specification.

DDC Performance Specification

Part I—General

Part II—Systems/Performance

Part III—Execution

Part I—General

The purpose of this section is to list general requirements for design and implementation. Large facilities with ongoing construction and renovation activities may provide site-specific instructions and requirements related to the long-term development of HVAC, BAS, and other systems. The section should also contain specific requirements for energy efficient design, references to energy conservation policies, references to established energy incentive programs, interaction with local utilities, and so on. Some sites have existing DDC systems and/or established long-term relationships, for example partnership agreements with preferred DDC vendors. Some sites may have developed in-house design review process, or other commissioning and TQM procedures.

All general requirements, site-specific conditions, and adherence to them should be included in this section of the specification.

Part I—General

Part I—General includes six sections:

A. Owner's Requirements

B. Instructions

C. Related Work by Others

D. Scope of Work

E. Submittal

F. Warranty & Response Time

A. Owner's Requirements

1. The Owner's project manager shall appoint a commissioning agent responsible for quality assurance of the design, implementation, and turnover. They jointly appoint a group of experts responsible for continuous review of the project at its predetermined phases.

2. The pre-selected DDC vendors shall review and bid on the HVAC design no later than at 75% design completion. This shall assure completion of the HVAC design with the selected DDC system. Changes after selection of the DDC system due to changes of the HVAC systems design shall be approved by the project manager and commissioning agent.

3. The design team shall give serious consideration to review comments, and if feasible, shall comply with the review comments. The design team shall also comply with site-specific standards and requirements pertaining

to site O&M requirements and existing practices. Possible discrepancies between the design, review comments, and site issues shall be resolved between the project manager and commissioning agent. This shall not relieve the design team from responsibility to design systems that meet applicable standards and code requirements, nor from associated liabilities.

4. The design team shall fully utilize available information and expertise related to the site, existing building systems, utilities, controls and automation systems, and other details offered by the owner's organizations.

5. The design shall comply with current site energy efficiency standards, and federal, state, and local utilities energy incentive programs.

Item A

The purpose of the Owner's Requirements section is to call the attention of the design team and DDC vendors to issues specific to the site and important to the owner. The previous section should list the owner's requirements pertaining to project organization, project review, quality assurance, and other site-specific requirements.

B. Instructions

1. This section shall specify work required **to provide a complete and fully operational controls and automation system** that interfaces with existing systems on site and is installed in accordance with current national, state, and site-specific standards.

2. The Engineer shall use this site specification in the job documentation, and add to it the following job-specific documentation:
   a. Sequence of Operation
   b. Point list with site acronyms
   c. Specific job related instructions
   d. Relevant structural, mechanical and electrical drawings
   e. Description of existing and new HVAC systems to be controlled, or interfaced to, by the control and automation system
   f. Related work under other divisions
   g. Energy conservation and rebate programs

3. The following DDC systems are approved for HVAC control applications:
   - System xxxxxxx by XXXXXXX
   - System yyyyyyy by YYYYYYY
   - System zzzzzzz by ZZZZZZZ

Item B.1

As indicated in the first paragraph of the Instructions section, the task is to provide a **complete and fully operational controls and automation system interfaced with existing systems on site, and installed in accordance with current national, state, and site-specific standards.** The words "complete and fully operational system" make the distinction between Requirement Specifications and Performance Specifications. The Performance Specifications put the responsibility for the proposed DDC system on the controls vendor responding to the specification. If, for whatever reasons (limited budget for the DDC system, HVAC engineering, inadequate capacities of connected HVAC systems, use of space, etc.), the DDC vendor cannot assume full responsibility for providing a "complete and fully operational system," the vendor should alert the owner's project manager and commissioning agent.

Although this request sounds unrealistic in competitive bidding situations, it is not so unrealistic for sites with a good working relationship with pre-selected DDC vendors. In fact, it is quite common in facilities that have long-term (working) partnerships with selected DDC vendors, have established the team approach to design and implementation, and who focus on the specified end result.

Item B.2

The second paragraph of the Instructions section specifies how the (HVAC) engineer is to organize the project documentation. A uniform organization of project documentation is important for sites with ongoing construction activities. The site specification should be presented in its original format for all projects, without modifying content or form. Job-specific requirements should be amended to the specification by the (HVAC) engineer. This saves time for the review team, which can focus on job-specific issues. It also saves time for the DDC vendors who already understand site-specific issues (from previous projects), and can focus on job-specific issues listed in the engineer's documentation. This is a real time saver considering the many different formats used by consulting engineers when they write and organize their specifications.

For sites with published and periodically revised standards, the project manager might ask the consulting engineer to refer to these standards in the job documentation, without copying the entire site DDC performance specification into the job documentation. Again, a real time saver for consulting engineers.

Item B.2.a

The Sequence of Operation sub-section of the Instructions section is a framework for the control vendor's application engineer to understand the HVAC engineer's intentions for systems operation. The Sequence of Operation should be a narrative description of the HVAC systems operation. It does not have to be an algorithm with defined control logic, parameters, and constants for (PID) control loops. Based on the engineer's sequence of operation, the applications engineer develops preliminary control algorithms, drawing on his or her experience and the DDC vendor's library of controls sequences from similar jobs. The application program should be reviewed and approved prior to being loaded into the DDC controllers and Operator Work Stations (OWS). Because every job is unique, the HVAC engineer has to develop a Sequence of Operation for every job. Schedules, set points, analog limits, and so on, should have preliminary definitions; and because they are all software-adjustable parameters, they can be finalized during the system's start-up and test operation.

The following is a list of the most common items which should be included in the Sequence of Operation, if applicable for the particular job:

1. General description of the HVAC system design, capacity, location, associated exhaust systems, coils, parameters, areas served, and so on.

2. Fan control sequence for occupied, standby, and unoccupied modes of operation, including:

   - Time schedule (daily, weekly, holidays, temporary schedules
   - Fan start/stop or speed control (one, two, or variable speed):
   - Interlocks with other systems (hardware or software interlocks to exhaust fans, minimum air dampers, fire alarm and other systems)

- Setback for unoccupied modes of operation and for summer/winter operation
- System restart after failure (what conditions should be met, which fans start first)
- Manual fan control (safety switches, three way switches)
- Fan feedback status (from a contact, flow switch, differential switch, current transformer, or feedback from a variable speed controller)

3. Temperature, humidity, and pressure control sequences for occupied, stand-by, and unoccupied modes of operation, for:
- Heating (preheat, reheat, heat recovery, zone heating)
- Cooling (mechanical cooling, free cooling, dehumidification)
- Humidification (control of humidifiers, steam generators)
- Pressure control of supply and exhaust air, room static pressure

4. Controls requirements and items that aid the systems engineer in definition of the DDC system:
- Interlocks of control loops (cascading of loops, interlocks between cooling and heating, air and radiation systems)
- Set points (preliminary values, calculation of set points based on measured variables)
- Sensors assigned for controlling and/or monitoring their locations
- Methods of making calculations (enthalpy, BTU, %RH, etc.), if used in the sequence
- Look up tables and charts, such as steam tables, psychometric charts, and others, if used in the software
- Accepted end-to-end tolerances
- Other relevant information

5. Zoning of the building and HVAC systems and their control for occupied, stand-by, and unoccupied modes of operation:
- Interlocks of perimeter radiation heating with HVAC systems
- Control of individual zones (based on lowest/highest/average space temperatures, combination of DDC control with local controllers or thermostatic control valves (TCVs))
- Required temperature/humidity/pressure parameters and tolerances for individual spaces, zones, ducts
- Other relevant information

6. Ventilation systems and their control for occupied, stand-by, and unoccupied modes of operation:
- Exhaust and fume hood control (pressure, volumetric, minimum fume hood face velocity)
- Interlocks with the building air supply, emergency procedures, and purge alarm of laboratory systems)
- Interfaces of application controllers such as fume hood controllers, Variable Speed Drives with the building DDC system via communications protocol or hardwired
- Responsibility for safety and air quality, response to alarms

7. Interfaces and interlocks with other building systems and their control for occupied, stand-by, and unoccupied modes of operation:

- Emergency shutdown and restart procedures
- Smoke alarm shutdown and restart procedures
- Hardwired safety protection
- Operating safeties, alarms, operator messages
- Other relevant information

Item B.2.b

Point lists with site acronyms (some vendors use entities, or objects) must be included in job specifications. Control vendors use point lists to size their DDC controllers, define field gear, and also to price hardware when estimating the controls job.

The point list as well as the entire project documentation should use naming conventions already in use by the facility. Each facility has unique names for its buildings, and specific naming conventions for HVAC systems. Names for HVAC systems may be as diverse as blowers, air-conditioning units, air-handling units, or VAV units. Uniform site naming conventions are also used by other computerized systems, such as maintenance management systems, accounting systems, and other facility-wide computerized systems. Using site-specific naming conventions is essential for the orderly implementation of systems, supporting documentation, and for uniformity among computerized systems installed at the site. Not following established naming conventions could result in additional work for the systems engineer if the owner requires interfaces to other computerized systems. Consulting engineers and contractors should use uniform point names throughout the design and installation, from the blueprints through the software and field labeling.

Without established naming conventions, the owner may end up with a building with three units called AHU-1, or with three air handling units named AC-1, AH-1 and AHU-1. This can happen with ongoing building renovations over several years, each designed by different engineering firms.

To correct point names in the application software, after it has been written and installed, is very time consuming. Point names are defined in point definition, and appear in every control logic, calculation, and alarm definition. In the absence of a "global" correction feature, changes of acronyms could take up considerable systems engineering time.

Develop naming conventions for new buildings with future expansions of HVAC and controls systems in mind. Future development of controls and automation systems is heading toward integration due to mass development of industry standard drivers, and open and standard communications protocols. Besides integration of HVAC controls systems, expect more and more integration of DDC controls systems into non-HVAC controls systems. Such systems may be utilities metering systems, power plant control and automation systems, maintenance management systems, and other facilities automation systems residing on the same network, and providing real time as well as management information to the users. Using uniform naming conventions throughout the facility assures greater flexibility for integration.

DDC vendors, consulting engineers, and individual facilities may have their "preferred" point list formats. The consulting engineer should consult the pro-

INPUT/OUTPUT SUMMARY TABLE

Building _____
System _____

Description

Inputs

Analog:
- Room temperature
- Duct temperature
- Water temperature
- O/A temperature
- Static pressure
- K.W.
- Flow

Binary:
- Flow switch
- Switch closure
- Static pressure
- End switch
- Smoke detection

Outputs

Analog:
- Modulating actuator
- E/P transducer

Binary:
- Control relay
- Solenoid
- Start/stop
- 2-Position damper mod.
- Smoke interlock

3-Point floating:
- Actuator
- Mercury relay
- Feed/bleed sav

Alarms
- High limit
- Low limit
- Run time
- Maintenance message

Features
- Scheduled start/stop
- Temperature setback/setup
- Optimized start (adapt.)
- demand limiting
- Duty cycle
- Ventilation delay
- Economizer cycle (DB)
- H/W reset w/OA temperature
- VAV control
- remote read/reset
- Tenant override
- Color graphics
- Point lockout
- Trend log

Exhibit 2-1. Example of a point list (ASHRAE guideline 1-1989)

ject manager or commissioning agent for the preferred format for site standardization.

Exhibit 2-1 is an example of a point list. This table is reprinted from the ASHRAE Guideline 1-1989.

To establish good site-specific naming conventions that can outlast several generations of systems, the facility should first create a "uniform building list." This should list individual buildings by name or by street address. Intermixing both may cause confusion, and possible problems with the "sort" features of computer programs (usually they sort by address, names, and so on).

To create uniform DDC acronyms, follow these basic rules:

1. Each acronym must be unique

2. The maximum length of the acronym should be equal or less than the shortest "name" field in the application software of any connected system. For example, if the shortest name field of one of the systems is eight digits, all acronyms should have a maximum length of eight digits.

3. The acronym should contain three levels of identification: building/system/point.

If the maximum length of the shortest name field is eight characters, the first two characters can be used for building name abbreviations, the next three digits for system identification, and the last three digits for point identification. Most of the systems on the market provide more than eight character fields for names. Some systems allow individual identification of all three levels, providing eight or more characters at each level. In addition, some systems provide an extended field of 24 or more characters for long descriptions.

Developing Acronyms

The longer the name field, the more descriptive the name can be. In the following example of acronym development, a relatively short eight character field was selected.

Shorten building names to minimum, but practical, abbreviations (e.g., Main Library–ML; Engineering building–EB; Research Center–RC; Building One–B1).

Make HVAC system names standardized and unique within each building. For example:

System Names	Abbreviations
HVAC systems with supply and return fans	Axx
Make-up systems without return fan	Mxx
Exhaust system without supply fans	Exx
Fume hoods	Hxx
Fan coil units	Fxx
Variable air volume boxes	Vxx
Booster fan	Bxx
Heat recovery system	Rxx
Hot water exchanger	Xxx
Steam system	Sxx
Hot water systems/zones	HWx
Condensate return systems	COx
Chill water system	CHx
Condenser water system	CWx
Domestic hot water system	DWx
Pneumatic system	PNx
Alarm panels	ALx
Chiller	CHx
Boilers	BOx

Make point names unique within each system. For example:

Point Names	Abbreviations
Supply fan start/stop	SFS
Supply fan feedback	SFB
Return fan start/stop	RFS
Return fan feedback	RFB
Pump start/stop	xPS
Pump feedback	xPB
Damper control	xDC
Damper feedback	xDB
Valve control	xVC
Valve feedback	xVB
Outside air temperature	OAT
Fan discharge temperature	FDT
Return air temperature	RAT
Mix air temperature	MAT
Pre-heat temperature	PHT
Hot deck temperature	HDT
Cold deck temperature	CDT
Room temperature	R—T
Room humidity	R—H
Duct humidity	D-H
Smoke alarm	SMK
Fire controls alarm	FCA
Fire damper	FD
Freeze alarm	FRZ
Filter alarm	FIL
Bearing alarm	BRG
Low system alarm	LO
High system alarm	HI
Summer/winter	S/W
Occupied/unoccupied	O/U
Enable/disable	E/D
On/off	O/F

When name fields are limited to eight characters, designate numbers 1–99 for equipment, with a long description in the extended fields, if provided with the system.

Examples of acronyms:
a. MLA01FDT: Main Library AHU #01 Fan Discharge Temperature
b. B1R1234T: Building 1 Room 1234 Temperature

Item B.2.c.

Job specific instructions, not covered in the standard site specification, should be included in the project specification with references to relevant sections of the site standard DDC specification.

Item B.2.d.

Relevant design drawings furnished by the HVAC engineer and/or controls vendor should be part of the bid and/or job documentation package.

Item B.2.e.

Descriptions of existing HVAC systems to be controlled or interfaced by the controls and automation system in retrofit installations belong to the HVAC specification (division 15000). However, it is listed here as a reminder to provide coordination and interlocks with existing HVAC controls systems serving the same or adjacent areas.

Especially for retrofit jobs, it is important to survey, document, and understand the existing building systems, their functions, operation, and maintenance problems, and to define the interaction of the new HVAC system with existing building systems. Many otherwise well-engineered projects fail because of the lack of coordination of newly installed HVAC systems with existing building systems (HVAC, utilities).

This is especially true if only parts of the building (a floor, a wing, etc.) are renovated. The air supplied by HVAC systems migrates from one (open) area (renovated) to another area (not renovated), and may create pressurization problems and an unpleasant, drafty environment. The situation may be created by lack of coordination of new HVAC and controls design with the existing conditions.

Pay similar attention to existing sensors and locations of the new ones, especially if they are located in the same or adjacent areas, or are being used in control algorithms. Other areas of interconnections that frequently cause problems when inadequately surveyed are building utilities systems, such as steam (failure to properly define pressure), condensate return systems (connection of low pressure condensate to existing high pressure lines), hot water systems (capacity of existing exchanger, zoning of distribution systems), and other existing building systems.

Related interfaces and controls coordination must be defined in the job documentation by the HVAC engineer for retrofit installations.

Item B.2.f.

Related work, and coordination of work with other divisions, should be described in the job documentation.

Item B.2.g.

For the last twenty years, every site and every building was touched by some energy policy and conservation program. In the last couple of years, besides governmental programs, utility companies became active in energy conservation, specifically in "demand side management." The idea is to encourage facilities to reduce energy consumption by designing, or retrofitting with, energy efficient systems and equipment. To encourage their customers, many utility companies now offer their customers energy incentives in the form of rebate programs. Energy rebate dollars, in combination with energy savings, reduce the payback period of a project to an acceptable level, thus encouraging owners to fund such projects. Each region, and each utility company have their specific energy conservation and rebate programs.

Because the programs are local, and many engineering firms and control vendors are national (or international), it is up to the project manager and commissioning agent to advise engineers and control vendors of the existence of such programs.

Some programs may require documentation of energy savings by trending certain parameters with the implemented BAS. The DDC applications engineer should be aware of such requirements, and set up the system with adequate history files, trends, and report generation capabilities. Also, compliance with site energy standards and policies should be part of the project documentation.

Item B.3.

The third paragraph of the Instructions lists the owner's preferred DDC vendors for the site. This advises the HVAC engineer to consider the pre-selected DDC systems, and instructs the project or construction management to solicit proposals from the pre-selected controls vendors only. If the site has existing control systems and the owner wants to continue using the same system (or new systems interoperable with the existing one from the same vendor), the instructions should state the owner's directions.

There are two contradictory requirements in DDC vendor selection: (1) The operations and maintenance department's desire to standardize on one or a limited number of DDC systems for the site, and (2) the project management and/or governing regulations' requirements for competitive bidding. Consider the motivation behind the two requirements.

First, consider the O&M perspective. It costs thousands of dollars for the O&M departments to train their technicians, establish hardware inventories, and become proficient in operating and maintaining the installed DDC system. While the cost of training and spare parts increases proportionally with each new system, the proficiency level for operation and maintenance decreases at a much faster rate. This is due to the complexity of each system, and of communications interfaces.

Each DDC vendor develops, and puts on the market, new systems, on average every five years. Considering that an average life span of installed systems is 15-20 years, and that each system can migrate to the previous generation, this means three to four new systems from each vendor in that time span. (There are numerous building automation systems with three to four tiers of migration, the oldest having been installed more than 20 years ago). Each system consists of an operating work station with associated programs, stand alone building controllers with associated control programs, and several application-specific controllers for each generation. Even an owner who standardizes on one DDC vendor may have, in fact, installed three, four, or more systems over the span of 20 years. Multiplying the potential number of systems by the number of vendors (pre-selected, or selected in competitive bidding), the O&M department can end up with an unmanageable number of DDC systems, software, and field hardware.

Second, consider the competitive bidding requirements. Competitive bidding may be required by regulations, or as a measure of cost containment. Bids are based on the provided (HVAC) documentation; the selection of systems on the lowest cost. The cost usually refers to first cost rather than Life Cycle Cost (LCC). Since bidding is administered individually for every job, a multiple building facility could end up with a variety of low cost incompatible DDC systems that cannot communicate with each other or with a common OWS. Operation and maintenance of such a facility could become very expensive and inefficient, and system upgrades could be very costly. The writer of the specification needs to understand the long-term goals for development of facilities automation systems, and communicate them clearly in the specification.

C. Related Work by Others

1. Electrical Work

2. Communications

3. Mechanical Work

1. Electrical Work

1.1 The following related electrical work should be provided by the electrical contractor:

a. Installation of adequate task lighting above each control panel, by adding light fixtures if necessary, regardless of whether specifically shown on the electrical drawings or not.

b. Provide color coding (e.g., blue) and labeling of controls components, such as junction boxes for control wiring, raceways, circuit breakers, and other components of the installation, to provide easy identification by the control technicians servicing the controls and automation system.

c. Install protective guards for circuit breakers supplying AC power to the control cabinets and control field equipment, in order to prevent accidental (manual) switching off of the circuit breakers designated for DDC.

d. Tag all power wiring and circuit breakers supplying AC power for the controls installation. Field controls panel and field equipment that require AC power should be marked on the job drawings.

1.2. For installations where the controls contractor is responsible for related electrical wiring, the controls contractor shall comply with all related code requirements.

1.3. For installations where the controls contractor provides design/build implementation of the DDC system, the controls contractor shall provide:

a. Connection of field sensors, DDC panels, and field controllers.

b. Connection of all DDC communication wiring.

c. Extension of dedicated 120 VAC power feeds to field devices (e.g., valves and damper actuators, VAV controllers, fan-coil controllers, etc.).

The classification of Building Automation Systems has taxed the conventional methods of organizing specifications. The organization of construction topics into Divisions and Sections is, in theory, done simply to identify and locate information. Contract documents usually explicitly state that the organization of specification sections is not intended to divide or assign the work among trades—this being under the purview of the contractor. In practice, however, specifications are generally written to address trade-specific issues within an individual section. BAS involve virtually all of the mechanical and electrical trades, while the DDC vendor is responsible for mounting and terminating proprietary devices, controllers, and implementation of the application software. It is the DDC vendor who ultimately relates the behavior of a staggering combination of sensors and actuators to control fans, dampers, valves, and pumps to make a space comfortable and efficient.

Installation of a BAS can be executed in several different ways. Large controls vendors maintain local branch offices that can provide comprehensive engineering and technical services. These vendors have sufficient resources to offer complete design/build installations of their DDC systems. They can be the sole DDC contractor for the job. DDC vendors with limited local resources have to subcontract parts of the job to independent contractors.

Job-specific requirements (in addition to the site requirements) related to other trades (mechanical, electrical, telecommunications, and other) should be defined in the job documentation.

Item C.1.1.a

Task lighting above DDC and other controls cabinets is often absent from electrical (Div. 16000) design. This may be due to the fact that at the time of electrical design the control cabinet locations were undetermined, the cabinets were later relocated, or the lighting fixtures were moved due to obstructions in the mechanical rooms. Whatever the reason, adequate task lighting is important for the controls technicians setting up or servicing the controls system. Task lighting should be adequate to read printed material, schematics, instruments, and prints on electronic components inside the controls cabinets. Task lighting should be approximately 50 footcandles. Practical experience shows sufficient lighting is provided by a two by four fixture, eight feet high, in front of wall-mounted DDC panels.

Item C.1.1.b

It is important to provide clear and visible markings of control components, identifying the controls hardware in mechanical rooms, tunnels, basements, and other poorly lit areas. Tracing down unmarked components in the maze of other distribution systems can be a very time consuming and frustrating task for the maintenance technicians.

Item C.1.1.c

Unprotected and unmarked circuit breakers are often turned on and off by building occupants trying to troubleshoot their electrical problems (tripped breakers). Such random testing of breakers may cause problems for the field computers and for the entire control system. Accidentally switching off the computers in the field cabinets will increase maintenance costs and the frustration of the operating and maintenance personnel. Protective guards installed on electrical breakers is cheap insurance!

Item C.1.1.d

Tagging wires (by the contractor providing electrical installation) will save time for wire testing and terminations. Wire tagging is also important for future maintenance and correlation with as-built documentation.

For installations where all electrical wiring is provided by the electrical contractor (under Division 16000), the controls contractor has to coordinate the installation of field hardware (such as control cabinets, field sensors), wire testing, and labeling with the building electrical contractor.

Item C.1.2

For some projects it is advantageous to separate the building electrical installation (specified in Div. 16000) from the controls-related electrical work. In such situations the controls contractor (and/or its electrical sub-contractor) is responsible for controls-related installation. This arrangement offers advantages for HVAC or controls retrofit jobs where the DDC contractor is contractually responsible for the entire (turnkey) job. It can be successfully used on larger jobs as well, provided there is a clear definition of responsibilities between the building and controls electrical contractors. Again, the controls contractor has full responsibility for all control wiring.

Item C.1.3

For controls design/build implementations managed by the owner, the owner's project manager and the commissioning agent take full responsibility for the design and implementation process. The DDC contractor is obliged to provide a fully operational system as specified in the site specification. The commissioning agent is responsible for quality assurance during the design, implementation, and systems turn-over to the O&M department. The advantage of such a direct approach, especially for retrofit installations, is in close relationships and teamwork among the facilities engineering and O&M departments, and the vendor's systems engineers.

2. Communications

  a. The controls contractor shall provide testing of new communication modes (other than the ones working in the facility) on the owner's premises and communications environment for compatibility with associated systems, speed of transmission, and error rate on the longest wire run. The testing shall run for at least one week. Results shall be submitted to the commissioning agent for evaluation prior to making a commitment for implementation.
  b. Feasibility and methods of communication to third party controllers (i.e., fume hood controllers, chiller controllers, boiler controllers, and other third party controllers) shall be determined jointly by the facilities engineer, design engineer, and the selected controls vendor for the job.

Item 2

In computerized and networked environments, error-free communication between systems is an important part of the implementation. Regardless of the organization of the job, requirements for communications (including wiring) should be defined jointly by the facilities engineers and the controls vendor. In the absence of a systems integrator, and in considering DDC communications only, the DDC vendor is fully responsible for the installed system's error-free communication.

There are several levels of communication:
  a. Controller to (front-end) Operator Work Stations
  b. Between master (intelligent building) controllers
  c. Between the master and connected application-specific controllers
  d. DDC to third party unitary controllers
  e. Communication to other systems utilizing drivers or protocols.

The communications used depend on the system architecture, vendor-specific controllers, and communication interfaces.

Physical media (communications wiring) used for DDC systems are dedicated twisted shielded pairs of wires, coaxial cables, fiber optics, and dedicated or dial-up telephone lines. Depending on the site, there may be a need for high-speed reliable communication over coaxial cables or fiber optics. For most applications, a twisted shielded pair or dedicated existing spare telephone wiring is sufficient. Due to high labor costs, using existing wiring or communicating over existing dedicated telephone lines is the most cost effective solution. However, prior to making this decision, the vendor should test the communication lines for signal clarity, strength, error rate, and reliability. There are many factors influencing data communications over a network. Some factors are site specific. The only way to find out whether the proposed communication will work at the site is to set up site testing for one or two weeks. Actual field conditions may require methods of improving communication signals, such as using modems, repeaters or hubs. The final communications proposal must include all components related to the actual site conditions.

Item 2.a

Control vendor's proposed communication methods other than the ones already in place at the site should be field verified.

Item 2.b

Requirements for building operations and associated environmental controls are getting more and more complex. In response to owners' requirements, numerous application-specific controllers have been developed over the past years. They are dedicated to control specific pieces of systems or applications such as fume hoods, chillers, boilers, rooftop units, and greenhouses. There are as many dedicated controllers on the market today as there are tasks to control. Most of these controllers were developed as stand-alone controllers, some with interfaces to their own dedicated front end PCs. This is seemingly a good idea, if you are looking at each individual system. However, their interface to a BAS system can be challenging. Before specifying such interfaces in the job documentation, be sure that the BAS vendor has developed an interface to the particular controller. Every BAS vendor publishes a list of developed interfaces to third party controllers. Verify that such an interface was developed, tested, and has been operational in a similar environment for an extended time, unless your site wants to be a test site for such an interface.

There is yet another challenge, this one related to information management. Two basic questions should be answered by the owner prior to deciding on networking:

a. What information do we *really* need, and where should it reside on the network?

b. What are we going to do with the information? (If you do not know how to use the information, you probably do not need it.)

Protocol or driver interfaces allow you to map over all points from third party controllers, but that may be more information than you bargained for. Because protocols and network interfaces cost money, the facilities engineer (on behalf of the O&M department) and the DDC systems engineer, together, should determine whether such interfaces are justifiable. Hard wiring certain points to the DDC system (necessary for proper control and reporting) can be a more cost effective and more efficient solution.

Another challenge is assignment of responsibilities for the reported data (namely alarms). An example of a possible logistical alarm problem that is also a nuisance is the interface to fume hood controllers. The local fume hood controller sounds an alarm (in the laboratory) any time the face velocity drops below a certain value, regardless of whether there is presence of toxic material in the hood. Such alarms could be reported to the facility's Operating Work Stations (OWS) via the interface protocol. While laboratory personnel may respond to this alarm routinely and either take corrective action or not depending on the situation in the laboratory, the DDC operator receiving the alarm must take corrective action. That can create an unnecessary additional load on the O&M department, especially if the facility has hundreds of such fume hood controllers and has to respond to each alarm.

If the established operating scenario warrants implementation of an interface protocol (between selected fume hood controllers and the DDC system), you have to decide:
   a. What points are to be shared by controllers in their operating algorithms?
   b. What information is needed at the laboratory level? (Assign responsibilities for responding to this information, i.e., laboratory safety-related information.)
   c. What information is needed at the central OWS? (Provide instructions for information management, i.e., who should respond to the alarm.)
The alarm problem demonstrates the importance of understanding the site operating philosophy, and a need for a facilities automation master plan. Operating and economical conditions should determine the extent of using third party controllers and protocols. Make the design team and systems engineer aware of such site requirements.

3. Mechanical Work

The following mechanical work should be provided by the mechanical contractor:
   a. Installation of thermal wells, hot taps, and other in-line equipment for sensors.
   b. Installation of control valves.
   c. Installation of outdoor dampers.
   d. Installation of other in-line devices, such as in-line meters, manual valves, gauges, and other equipment as specified by the HVAC engineer in the site documentation.

Item 3

This is usually part of the mechanical contractor's work, under supervision of the controls contractors.

D. Scope of Work

Provide labor, material, services, equipment, and the engineering necessary for a complete and operational Controls and Automation System, as indicated in the job-related documentation and specified herein, including but NOT limited to the following items:
   1. Complete systems engineering, with a sequence of operation, software generation, its testing and loading into the controllers. development of colorgraphics for the Operator's Work Stations (OWS), development of engineering documentation, operation and maintenance manuals, and other documents as further specified in the Submittal and Execution sections of this specification.

Communication   Levels

Item	Site	System	Building	Controller	3rd party
Communications mode					
Communications media					
Redundancy					
Modems, hubs, etc.					
Transmission speed					

2. Project management from design through implementation, testing, and turn-over of the system to the owner.

3. Furnishing and installation of the specified control system including supervision of installation of field devices such as dampers, valves, wells, taps, wiring, etc., as specified in the job documentation, and installed by the mechanical and electrical contractors under Divisions 15000 and 16000.

4. Control of air handling systems as specified in the job documentation.

5. Control of hydronic systems as specified in the job documentation.

6. Control of other related equipment, such as boilers, chillers, compressed air, secondary reporting of fire alarm systems, and other systems connected to the controls and automation system, as specified in the job documentation.

7. Site-specific scope of work for DDC communications: The controls and automation vendor shall review the following communication requirements with the owner's engineer for each level of communication:
   a. Communications mode—lower level protocols (Ethernet, Arcnet, RS-485)
   b. Communications protocol (vendor specific, Modbus, BacNet)
   c. Communications media (twisted pair, coax cable, fiber optics, dedicated existing telephone pairs, etc.)
   d. Redundancy (dual trunk, dual hubs)
   e. Use of modems, hubs, multiplexers (their location on the network, purpose, speed)
   f. Transmission speed (9600 baud, 19.6kbps, 2.5Mbps, 10Mbps)

8. Based on this site-specific scope, the controls and automation vendor shall submit with the proposal a table showing the communication characteristics of the proposed system.

9. The controls and automation vendor shall review the following information requirements with the owner's engineer, for each level of communication:
   a. Point mapping (which points should be transmitted over the network from one network level to the next level)
   b. What is the purpose of the information and how to use it
   c. Which points are to be used in control algorithms by which controller?
   d. Which points are to be used for alarm reporting?
   e. Which points are to be displayed in the colorgraphics at the Operator's Work Station
   f. Instructions to the operators (for example, how to use the non-HVAC related information)
   g. Instructions on how to access information from other systems, and how to provide automatic transfer in real time or batch file mode (how to log off and on to different systems from the same terminal, how to transfer

an alarm from an HVAC control system to a maintenance management system to print out a work order, etc.)

Based on this review, the controls and automation vendor shall submit with his proposal a draft related to communications within the scope of the proposed system.

Item D.1

For performance-based jobs, the DDC vendor is responsible for engineering of the DDC system—to deliver a completed and fully operational system to the owner. Systems engineering, operating logic based on the reviewed and approved Sequence of Operation, and generation of application software are the most important engineering activities that assure proper operation of the environmental control systems. Person Machine Interfaces (PMI), including colorgraphic operator screens, alarm reporting, and management report generation, are essential tools for the operators. DDC systems documentation and requested manuals are necessary to operate and service complex environmental control systems.

Item D.2

Most control vendors assign a project manager for systems implementation. The owner's project manager and commissioning agent should work with the vendors project manager throughout the systems implementation process.

Item D.3

Requirements for the furnishing and installation of the DDC system's components is specified in the job documentation.

Items D.4 through D.6

Scope of work for the job is specified in the job documentation. Items in the specification list the areas most common for DDC jobs.

Item D.7

Site-specific requirements related to communications should be included in this section of the specification. Controls and automation vendors design their systems with different communications interfaces and protocols (i.e., RS-232, Arcnet, etc.). Most DDC systems can communicate over a variety of physical media, such as twisted shielded pair, dedicated telephone lines, coax cables, and fiber optics. Because every site is different, site communications requirements must be defined by the owner's engineering organization. If the site has an already established (and working) network, specify the same site communications requirements for the DDC system. If the site has no preferred communications architecture, the facilities engineer, telecommunications specialist, and an independent systems integrator should evaluate the situation and develop a plan for the most suitable network for the site. However, in the absence of standard communications protocol, a plan cannot be developed without an understanding of the communications options of the BAS considered for the job.

Figure 2-2 is an example of a communications scenario with multiple HVAC controls and automation systems designed with open (not standard) communication protocols. Communication levels or tiers shown on the drawing segment the information in relation to its characteristics: speed, volume, and type of information. For large networks, such architecture also provides

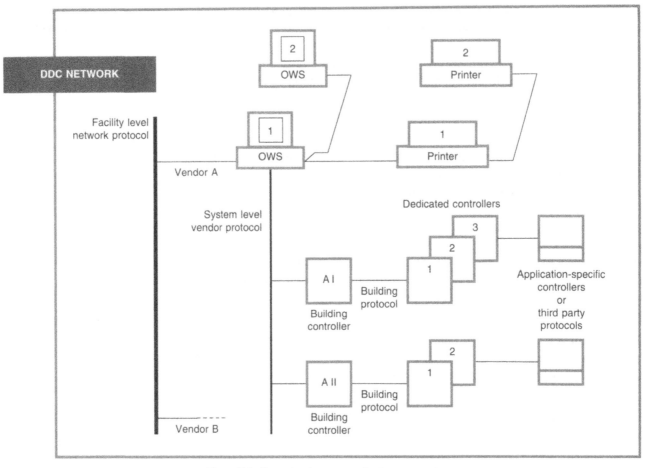

Figure 2-2. Example of a communications scenario

compartmentalization, in case of errors which may affect an entire segment of the network.

- At the site level, the automation server can communicate with other facilities' computerized systems to share *management information* via industry standard protocols.

- At the system level, DDC systems of different manufacturers can communicate with each other or with a designated server via a common or dedicated protocol.

- At the building level, system controllers located in each building can communicate with each other and with their designated operator interfaces via open or proprietary protocols.

- At the controller level, dedicated controllers communicate with each other and with building controllers via open or proprietary protocols.

There are numerous combinations of network architecture, network topology, utilized protocols, and drivers. To design a network that best fits the site requirements and is in compliance with communications options of the automation systems on the market, larger sites need to develop a site communications master plan.

One practical recommendation for networking of existing systems: *implement one interface at a time; completely "debug" the system before moving to the next interface.*

The owner's facilities engineers along with the participating vendor's systems engineers have to define communications related items at each level of communication. The DDC vendor responding to the specification has to list all communications-related information (at each level) in the form of a table for the facilities engineers or systems integrators to evaluate.

Item D.9

Networking and information transfer is becoming more and more important in facilities automation in general, and in DDC systems in particular. The traditional approach of transferring one signal (on/off) on a pair of wires cannot be applied in the world of networking and communication protocols. Control vendors are valuable resources for facility engineers in searching out the best communications scheme for their systems. However, interfacing to third party systems over non-proprietary protocols requires bringing the third party's systems engineer to the table. It is prudent to use both vendors' expertise to avoid pitfalls and costly delays during the start-up and future systems operation. Do not try to rely on only one vendor's expertise.

Networking requirements of large facilities that have a variety of existing controls and BAS is a real challenge. For such facilities, development of a site communications master plan is essential. However, its development should be an independent undertaking. References to site networks and site master plans should be part of the site specification, especially if the controls and automation system is to be connected to the site network.

E. Submittal

1. Pre-implementation Submittal

2. Implementation Submittal

3. Turn-over Submittal

1. Pre-implementation Submittal

The controls and automation vendor shall deliver to the owner's project manager four hard copies of submittal documentation as described in this section. The documentation shall be delivered on 3.5 inch IBM-formatted diskettes using the following software: the drawings in AutoCad®, Rev. x or higher; tables in Lotus 123 spread-sheet program, Rev. x or higher; and reports in WordPerfect for Windows word processing program, Rev. x or higher.

   a. The preliminary submittal shall be delivered with the final proposal of the BAS, and will include the following:
   1. Systems description
   2. System architecture
   3. Point list
   4. Controls diagrams
   5. Cut-sheets of system hardware components
   6. Description of communication interfaces including communication test results on the owner's premises
   7. Other job-specific items as requested in the job documentation
   8. Final price for the contract, including all discounts and mark-ups

9. Detailed and guaranteed job schedule

b. The following items shall be delivered during the design and engineering phase, and at different stages of completion, for QA review and approval:

1. Systems engineering documentation such as (a) detailed system description, (b) detailed systems architecture, and (c) detailed design of communication modes and protocols.

2. Final Sequence of Operation, detailed controls algorithms presented in the form of generic flowcharts, Graphic Programming Language (GPL), or line programming, if utilized by the system. The documentation shall also include calculations, tables, equations, and other supporting information used in the application program. These requirements apply also to optimization, simulation, and other advanced application programs for the job.

3. Drawings with control diagrams, including hard wired interlocks and interfaces to other systems via third party protocols. The diagrams shall use site-specific acronyms.

4. Drawings of colorgraphics developed for Operator Work Stations (OWS).

5. Detailed and final point list, including names, acronyms, field hardware types, models, manufacturers, and locations. Sensor ranges, means of calibration, and initial values for control and set point variables shall also be included in the documentation.

6. Drawings of cable plans, riser diagrams, locations and field panel layouts, locations of field points which require AC wiring, and termination and wiring diagrams of all control points.

7. Drawings showing location of field hardware installed by the mechanical contractor, such as thermal wells, and valves.

8. Copies of individual product data sheets.

9. Sample copies of standard reports, instructions for user generated reports, and samples of trend and history data reports in tabular and graphical format.

10. Tabular presentation of definition and mapping of points transmitted over a communication protocol; network layout, network security, and access levels for each node on the network.

11. A set of engineering documentation containing spring ranges, pneumatic pressures, and requirements for pneumatic calibration for pneumatic devices.

c. The following submittal shall be requested prior to beginning the field installation:

1. A set of test results (functional, simulation, interfaces, etc.) and/or observation of factory tests.

2. A hard copy of the final application software for each control panel or computer containing such software, including operating logic, associated definitions, and interfaces.

Item E

Owners should be specific about submittal requirements from the controls and automation vendors. In this way, the owner establishes uniform requirements for the structure of submitted documentation, and provides checkpoints for turn-over of individual submittals to the commissioning agent. This makes the project management more straightforward, eliminating second guessing and future disputes about submittals, their contents, and completion dates. It also aids the operation and maintenance groups with standardized site documentation, manuals, and reports.

The submittal section of the contract documentation should specify dates for review of submittal by the commissioning agent and review team.

Deliver all submittals to the project manager and commissioning agent in a hard copy format (printed on paper), as well as on diskettes compatible with the owner's word processing software, drafting software, spreadsheets, etc. The contract documentation should specify the number of copies required, and any software-related requirements. With the enhancement of networks, you can easily transfer text files (Lotus, WordPerfect, etc.) and graphic files (AutoCad or its conversion to .DXF files) via the network.

Item E.1.a

For projects where the controls contractor is a sub-contractor to a mechanical contractor or construction manager, the requested items under Item E.1.a. become part of the DDC vendor's bid documentation. For design build projects, or when the controls vendor is contracted directly by the owner, the documentation should be delivered to the owner's project manager. In either situation, deliver the documentation to the commissioning agent for a quality assurance review.

Item E.1.b

The submittals in Item E.1.b. are important for continuous QA review during the systems engineering phase. In this phase, all systems engineering and application software development, including sequences of operations, should be defined, developed, and the programs written. The intention of continuous QA review is to focus on the operating logic, interfaces, and interlocks, and to approve them prior to generating the software. Undetected problems related to operating logic and interfaces will cause future performance problems. Early detection of such problems is the least expensive way to their elimination.

Item E.1.c

The owner's commissioning agent may request to be present during the factory tests of the controls and automation system, prior to its shipment to the job site. Factory testing is not as much a hardware test as a test for the application software, interlocks, communication protocols, point mapping via third party protocols, and other job-specific items. It is assumed that prior to loading the application software the vendor did Quality Assurance (QA) testing of the associated hardware and operating software.

If the job uses a standard off-the-shelf DDC system already in operation on several (similar) job sites, the owner can waive factory testing requirements. In that case, it is advisable that the commissioning agent ask for a demonstration of some procedures, such as system setup, software loading, downloading of the application software to the controller, uploading of the points to the server, and other procedures, as part of ongoing QA. Alarm definitions should be specified and limits defined before systems start-up. Such precautions can save boxes of printer paper during start up. If alarms are not properly defined, clearing up the nuisance alarms may take days or weeks. The same is true of operating schedules and initial set point values for the control loops.

2. Implementation Submittal

Upon completion of the job and prior to final turn-over, the following submittal shall be provided by the controls vendor:

a. Control vendor's standard hardware and software manuals, and cut sheets of third party components.

b. Field test results, calibration and verification lists corrected for the final conditions.

c. As-built final shop drawings, including floor plans, control schematics, field gear locations, wiring diagrams, control cabinet layouts, and other drawings pertaining to the installation.

d. Instruction manual(s) with functional descriptions of the controls and automation system and its components:

   (1) Description of operation, including sequence of operation under normal conditions, system start-up/shut down, and operations under emergency conditions

   (2) Fail-safe conditions of controlled devices and equipment

   (3) Instructions for system maintenance; a list of all devices, including their identification and location, accessibility, service requirements, and frequency of service; normal positions of controlled devices and equipment; final ranges and control settings; calibration requirements and frequency; benchmark values; special test equipment; and other maintenance-specific instructions.

e. Application software manual containing:

   (1) Overview of the application

   (2) Detailed description of the operating logic accompanied by line or graphical programming, and a definition of hardware and software points

   (3) Instructions for the operators regarding initiating and disabling programs, and generating reports and trends; procedures for uploading/downloading of programs; troubleshooting software problems; and relevant information for periodical software maintenance.

   (4) Instructions for user programming, mathematical calculations, modifications of the controls program, and changes of control variables.

f. As-built communications manual, including:

   (1) Drawings of communications layout with hardware, nodes on the network, and modes of communication

   (2) Manufacturers' cut sheets of all components

   (3) Drawings of wiring with circuit numbers, polarities, baud rates, jumper settings, etc.

   (4) Description of communication protocols

   (5) Point mapping

   (6) Relevant engineering documentation for third party systems.

The documentation and manuals are part of the final job documentation. The documentation should be reviewed by the commissioning agent as part of the ongoing review process, updated during the systems start-up, and turned over to the owner upon job completion.

3. Turn-over Submittal

The following final submittal shall be turned over to the owner after completion of the test operation and prior to the beginning of the warranty period:

1. All documentation, corrected for as-built conditions
2. Field test protocols
3. Sign-off sheets

Final turnover of documentation to the commissioning agent should be part of the contract. As-built documentation and operating and maintenance manuals are essential for systems operation and maintenance. As-built documentation should contain sensor locations, building zoning (indication of which rooms are served by what air handling unit or heating zone), equipment location (especially in hard-to-find places such as basements, above drop-ceilings, and so on). Field test protocols and other information related to initial values of calibration, set points, constants, etc., are essential data for follow-up services.

F. Warranty and Response Time

During the warranty period, the controls and automation vendor shall guarantee the proper operation of the control system and implemented control strategies. During the warranty period, the controls and automation systems vendor, HVAC engineer, and the owner's operations and engineering organizations examine and verify the performance of the controlled HVAC system. Any tuning, software corrections, or enhancements to the application software algorithms installed within the scope of the project, and aimed at meeting the building design parameters continuously during different seasons, shall be made at the request of the owner in consultation with the engineer, and provided by the controls and automation vendor at no additional cost to the owner. However, this software warranty does not include changes due to building use, building occupancy, design changes, changes of design parameters and criteria, and changes of HVAC systems configuration. It does not entitle the owner to free software releases unless otherwise specified in the contract documentation.

System engineers and/or service technicians shall be dispatched to affect remedial action within four (4) hours of being notified. Service shall be available year round, 24 hours a day.

Automation systems are complex, controlling intricate mechanical and HVAC systems to provide healthy and safe working environments in the buildings. Unlike hardware installations, automation systems warranties should relate to systems performance rather than to the materials and labor of installed systems. Application software, operating logic, and other items defined in the software are tailored to control the installed HVAC systems. The performance warranty of a DDC controls and automation system should include hardware, software, operating logic, communications, and other items important for operation.

There are several warranties offered by controls and automation vendors:

- *The basic one year warranty* (some vendors have extended their basic warranty to two or more years) is usually included in the proposal for no additional charge.

• *The extended warranty* may be requested by the owner in the job or standard site specification, and is provided by the vendor for an additional cost.

Most requirement specifications phrase the warranty requirement as follows: "The system shall be free from defects in workmanship and material under normal use and service for a period of one year after system acceptance. If, within this guarantee period, any defects in materials or workmanship occur, remedial action shall be provided by the controls and automation vendor to repair or replace any defective work without additional cost to the owner."

This warranty applies to workmanship and material, not to systems performance. The warranty statement addresses hardware defects, but an installed DDC system may be without "defects in workmanship and material," yet unable to maintain the design conditions continuously, year round. Obviously, such a system does not meet the owner's requirements for occupancy comfort.

A performance specification requires the vendor "to provide a complete and fully operational controls and automation system." Warranty requirements should focus on system performance in addition to warranties on material and workmanship. These requirements should be understood at the project definition phase, and money for a performance warranty should be allocated in the project budget. Such action would allow the owner to operate the environmental control systems under real-life conditions and seasonal changes, and to make modifications to the operating logic when unforeseen problems surface. A performance warranty eliminates lengthy discussions (even legal actions) on who is responsible for the problem, and lets the engineers focus on improving the systems operation. Explaining the performance warranty conditions to the building occupants, and asking them for their input makes them part of the system start-up. Their cooperation is invaluable for the transition phase, from start-up to full occupancy.

Part II—Systems Performance Specification

Development of automation systems is influenced by the development of field hardware, such as sensors and actuators, and by the development of computer technology. New developments in computer hardware, operating systems, networking, and other areas have an impact on an automation system's architecture, application programs, interfaces, communications, and other areas vital for a system's performance. A major leap in systems development took place with the introduction of distributed Direct Digital Control (DDC) systems, which replaced the centralized Energy Management Systems (EMS). This change generated a chain reaction in systems definition, architecture, operating and application programs, and also in specifying controls and automation systems.

Systems specifications written by HVAC (specifying) engineers often describe systems characteristics common to every system of the same generation on the market. Because all off-the-shelf DDC systems on the market have similar components and characteristics, systems specifications should focus on descriptions of requirements specific for the site or application. Specifications do not have to state obvious features and characteristics standard in every DDC system. However, they should describe site- or job-specific requirements, and match them with the features of standard off-the-shelf systems. System customization, other than development of application software and utilization of available options of a "standard" DDC system, is not recommended. Modification and customization of standard off-the-shelf systems architecture, operating software, or modes of communication could be a very expensive undertaking.

Systems specifications should be updated frequently to reflect the latest generation of DDC systems on the market. The practice of getting by with one specification for several years is gone, due to rapid systems development. Because of ever-evolving technological changes, "standard" specifications should be revised and updated at least annually, to keep up with the latest systems development.

Part II—Systems Performance Specification

Part II—Systems/Performance includes two subsections:
   A. System-specific Performance Requirements
   B. Site-specific Performance Requirements

A. System-specific Performance Requirements

1. Distributed Processing

DDC processing shall be performed at the lowest microprocessor-based controller level. Application software, including all parameters and set points, shall reside in the lowest level controller designated for standalone operation. The entire database of this controller shall be accessible either from the front-end OWS, local PC, or hand-held terminal.

Standalone capabilities shall include:
   a. Closed loop control functions (P, PI, PID, Incremental, Floating, etc.), cascading of control loops, tuning of control loops, and on-line calibration of connected analog field devices.
   b. Execution of standard energy management functions
   c. Execution of custom written program.
   d. Alarm reporting.
   e. History data storage.

Depending on the system, processing can be performed at various levels of the DDC architecture. The most common approach is to provide processing at the building DDC controller level. In such cases, point definition, control logic, processing and data storage, as well as communications interfaces reside in the building controller. The building DDC controller is capable of processing all control and defined energy management functions as a standalone unit, and also provides for display and local access via a portable PC or hand-held terminal. Some systems offer processing at the unitary or application specific controller level, or a combination of processing and data storage between the building controller and field (unitary, or application-specific) controllers.

Figure 2-3 is an example of a Johnson Controls, Inc. Metasys software architecture. Note the features processed in the Network Control Module (NCM), which can be considered as a building controller, versus the features in the Digital Control Module (DCM), or Extension Module (XM). Because each DDC software architecture is different, understanding it is essential for system specification and selection.

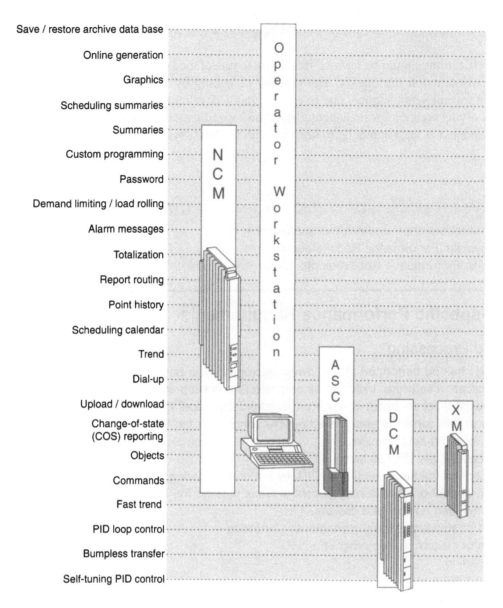

Save / restore archive data base
Online generation
Graphics
Scheduling summaries
Summaries
Custom programming
Password
Demand limiting / load rolling
Alarm messages
Totalization
Report routing
Point history
Scheduling calendar
Trend
Dial-up
Upload / download
Change-of-state (COS) reporting
Objects
Commands
Fast trend
PID loop control
Bumpless transfer
Self-tuning PID control

Figure 2-3. Metasys Software Architecture (Johnson Controls, Inc.)

These requirements should be clearly stated in the site specification, and confirmed by the DDC vendors bidding for the job.

A remote standalone DDC (i.e., building controller) should perform its full control and energy management functions, regardless of the condition of the communications link with the front-end PC or other computer.

Controls and automation systems (and their standalone controllers) should be designed with maximum fault tolerance, such that a failure of one controller will not disrupt communications to the Operators Work Station (OWS) or other controllers on the controls and automation network.

2. Energy Management

The following energy management programs shall be provided with the DDC system:

    a. Time scheduling
    b. Supply air reset

c. Economizer control (dry bulb), also called free cooling
d. Optimal start/stop using adaptive algorithm
e. Chiller optimization
f. Facility and system-specific programs

These energy management programs are the most common supplied with DDC systems. They were developed to optimize HVAC and building systems operation, and to provide energy savings to the customer. Because every job is different, the commissioning agent, HVAC engineer, and the vendor's systems engineer should determine which energy management programs will reduce energy consumption without inconveniencing the building occupants.

Item A.2.a

Time programs or weekly scheduling of HVAC systems operation is the most prominent energy conservation program. A number of start and stop programs can be pre-programmed for controlled equipment (fans, pumps) for any given day of the week, for holidays, and for other special schedules. Time programs can also work with reset programs that reset temperatures, air velocities, and so on, for pre-programmed time windows.

Item A.2.b

Supply-air reset programs, using space load demand sensed by either space sensors or a return air sensor, reset the set point of supply air to maintain desired space temperatures and are very effective energy saving programs.

Item A.2.c

Economizer programs, also called "free cooling," are based on the outside air dry bulb temperature. The program saves energy by using outside air rather than mechanical energy to maintain desired space conditions.

Item A.2.d

Optimal start/stop program using adaptive algorithms saves energy by adjusting optimum start and stop time of AHUs based on the space load and outside air temperature.

Note: The optimum start/stop program works well, providing: (1) building thermal loads and HVAC systems parameters are calculated fairly accurately, and (2) the central heating plant can satisfy multiple load start-ups (controlled and uncontrolled) at the same time. An adoptive control algorithm can take care of the building load dynamics.

The second concern is more serious in large plants with multiple boilers and a combination of BAS controlled and locally controlled loads (i.e., HVAC systems controlled by BAS are controlled loads; domestic hot water heaters, process steam, and locally controlled radiation heating systems, are considered locally controlled loads). The problem in using optimum start programs is in the random (optimum) start times calculated for each load. Because the calculated optimum start times are not coordinated with other controlled and uncontrolled loads, bringing on-line an "unpredicted" large load can cause temporary plant overload. This may be caused by insufficient on-line boiler capacity during winter morning start-ups. Keeping large boilers on-line, even at an idle firing rate, in order to meet momentary unpredicted demands, costs hundreds of dollars an hour. Such expense may be greater than the perceived savings achieved by optimum start programs not coordinated with the loading of the heating plant. Unless the optimum start/stop program can advise the heating plant on the forecasted loads, the program should be used selectively.

Item A.2.e

Most chiller optimization programs relate to (1) chiller and/or pump sequencing, (2) chilled water temperature optimization, and (3) condenser water temperature optimization, or a combination of them.

Item A.2.f

Some facilities may request special optimization programs which have to be custom written for their application. Most DDC systems on the market allow either customer programming, or some degree of modification of standard energy management programs.

3. Standard Control Routines

All programming shall be either from the Operator Work Station on the network, or from the local building controller via its hand-held terminal or portable PC.

    a. Programming routines available with the DDC system shall include, but not be limited to, the following:

- Math Routines—basic arithmetic, Boolean logic, relational logic, and fixed formulas for psychometric calculations.
- Utility Routines—process entry and exit, variable adjustments and output, alarm indication, power fail restart, local I/O interface.
- Control Routines—signal compensation, loop control, cascading, energy conservation routines, time programming.

    b. The following library of control routines shall be available in, but not limited to, the firmware, and capable of generating additional programs for specific controls requirements: "Dead band" control, psychometric chart DX control, totalization of historical data, advanced scheduling and holiday programming, lookup tables for calculations, and so on.

    c. Other control specific requirements:

      1. Final control programs shall be stored in EEPROM or battery backed-up RAM for up to 72 hours at power failure. In addition, all software programs shall be stored at the Operator Work Station computer's hard disk drive for review, modifications, and downloading to the building controllers.

      2. To maintain long-term analog accuracy in the analog sensing circuits, the controller's firmware shall compensate for term drift or drift due to ambient temperature changes.

      3. The remote building DDC controller shall contain in its program a non-destructive self-test procedure for testing memory and processor functions. Alarms shall be sent to the Operator Work Station for any abnormality within the processor and the memory, for all field analog point failures, and for all program deviation. Process and hardware variables shall be identified as being reliable or unreliable. When a calculation is required to use a value, sensed or calculated, that is identified as unreliable, the calculation will use a preprogrammed default value (the default value must be defined by the owner), and the unreliable status shall be reported at the Operator Work Station(s).

      4. Sufficient redundancy shall be provided that operation can continue

unimpaired given the failure of any single building control unit. Failure of any point module shall not impair functions of other modules, control loops, controller, controls and automation network, Operator Work Stations, and other components of the system.

The building DDC controller should contain all necessary mathematics, logic, utility functions, standard energy calculations, and control functions in ROM to be available in any combination for programming.

Introduction to Site-specific Performance Requirements

While system-specific items and DDC features are common to most systems on the market, site-specific requirements are unique for each site. The owner's facilities engineers are in the best position to describe site-specific issues in the performance specification. The following items are a collection of the most common site requirements.

Introduction to System Architecture

DDC system architecture is in constant evolution. Development of distributed architecture was a major leap from centralized BAS systems. This was a result of advancements in computers and in distributed processing. Today, systems architectures are most influenced by developments in the fields of communications and networking. Building controls and automation architecture went from central to distributed processing, and to more distributed processing at the equipment level. Advancements in communications and development of protocols will eventually erase the different levels of the current architectures, each controller and front-end PC becoming a node on a network.

Deciding on the DDC systems architecture, facilities should consider their current needs as well as future requirements for expansion, systems migration, facilities integration, and networking. It should always begin with the technology currently on the market. One cannot, and should not, wait for future developments because they too will be surpassed by newer developments. However, facilities with ongoing construction activities should implement systems which can be integrated into future systems and networks. DDC systems architectures should be in accord with the long-range development plans of the facility.

While new development in communications and protocols will undoubtedly change the overall system architecture in the future, the classic three levels—front end, building controllers, and application-specific controllers—will most likely remain. The three-level architecture is a reflection of the physical layout of buildings and systems (HVAC systems located in buildings monitored by local or remote OWS). Use of vendor-specific systems for individual buildings also provides compartmentalization of individual buildings and vendor's systems. Such compartmentalization is appealing for systems troubleshooting, vendor support, systems operation, hardware maintenance, and for overall systems integrity.

Figure 2-4 is an example of three-level systems architecture. A site description of systems architecture should be included in this section of the specification.

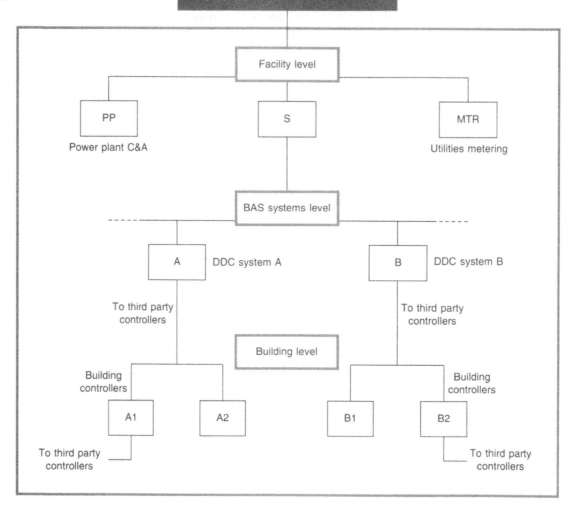

Figure 2-4. Example of three-level systems architecture

B. Site-specific Performance Requirements

1. System Architecture

a. PC-based Operator Work Stations (OWS)

The DDC vendor shall provide the following: Operator Work Stations—X number of front-end PC OWS located in the control room shall be connected to the network allowing simultaneous data access by the operators. One of the PCs shall contain the archive data base and all history files. The PCs shall be sized for the total capacity of up to xx,xxx field hardware points, xxx colorgraphic screens, xxx trend graphs, and for history data storage capacity of xxMB. Analog and digital values on the screens shall be dynamically updated each xx seconds; alarms shall be reported at the same interval. History files shall be updated by x minute averages from scanned and calculated points.

The Persons Machine Interface (PMI) shall allow the operators to move from one colorgraphic screen to another one in a logical sequence, using a minimum of mouse clicks or key strokes. The PMI shall allow the operators to acknowledge alarms, control devices, change set points, initiate printouts, access files, and so on, from the colorgraphic screens.

The password protection shall allow for grouping of points, grouping of buildings and grouping of responsibilities for at least xx access levels.

Printers connected to the front-end OWS shall be silent color printers with ASCII character set. One printer shall be designated as the alarm printer.

b. Building Intelligent Remote DDC Controllers

Building controllers shall be standalone, modular intelligent DDC controllers containing the database for the entire building. Each controller shall have a xx% spare point and xx% spare space capacity, or capability for expansion, to accommodate future expansion. Each controller's database shall be accessible via hand-held terminals or portable PCs only. Building controllers shall (or shall not) contain local displays and keyboards on the face of the panel. Access shall be password protected to prevent accidental access of database in other buildings. A higher access level shall allow the DDC technicians to access other controllers on the network.

c. Application-Specific Controllers

Application-specific controllers shall be specified in the job documentation.

d. Approved Protocols and Information Transfer

1. The following protocols are approved for the facility:
   - At the facility level
   - At the BAS level
   - At the building level
2. Safety related points from one controller utilized in control algorithms of another controller shall be hard wired.
3. Only system alarms shall be centrally monitored from local controllers in areas that are fully staffed (e.g., power plants, laboratories, greenhouses). Application-specific controllers operating in areas without staff supervision (e.g., lighting controllers, individual chiller controllers, dedicated controllers for critical areas) shall be interfaced to the DDC system via protocols.

Item B.1.a

Individual OWS interface with their respective building DDC controllers via open or vendor-specific protocols. The specification should describe overall application requirements for OWS computers (such as memory capacity, and expected total number of colorgraphic screens), the number of OWS in the system, their use, and location. The specification does not have to provide detailed specifications of the type and make of the OWS PCs, unless the site has a purchasing policy in effect. DDC vendors have spent several man-years designing and testing different front-end computers and operating software for their DDC application software. The owner should follow the selected DDC vendor's recommendation for front-end hardware. If the front end is already installed, the site specification should define the interface requirements.

Item B.1.b

Building intelligent remote DDC controllers provide distributed processing, interface to the network or front-end OWS, and interface to application-specific dedicated controllers. The controllers should be located in the highest concentration of field points and have sufficient point modules for field point connections. The specification should define the spare capacity and spare I/O modules requested for the site. For modular DDC architecture with a set number of point modules, the expansion capability of the system should be defined. The specification should also state requirements for communications to the network or front-end OWS. Some sites may require a display and keyboard on the controller's front panel; others require access by a hand-held terminal or portable PC. Some sites may require control panels with locks and a door alarm that reports to the front end OWS. Such security measures should be defined in the specification.

With the advancement of networking and information distribution, more and more facilities require remote monitoring of the building functions by tenants or departments occupying the building. The site specification should define the number of local OWS, access levels, and the overall integration issues in the site specification.

Item B.1.c

Application specific controllers are used more and more with specific equipment such as fume hoods, elevators, lighting, and other building systems commonly found in modern buildings. Even though some of the equipment is not related to environmental controls or energy management, the convenience of being able to monitor them via the controls and automation system makes these interfaces desirable.

Which controllers should have a protocol interface, which should be hardwired to the DDC controller, or which should have a combination of the two is an application rather than a technological decision (provided there is an interface available). Cost and operating requirements should be the determining factors.

Specification of application-specific controllers should be in the job documentation.

Item B.1.d

Approved protocols and information-transfer-related issues should be described in the specification. In the absence of one standard communications protocol, the owner's engineers or systems integrators have to specify the acceptable communications protocol for each level of communication. The owner should also specify information-related issues, such as who is responsible for information transfer, its utilization by computers or humans, and processing of information (such as controlling or initiating response to alarms).

2. Modes of Communication

Facilities Network communication shall be via fiber optic cables, utilizing xxx protocol at a communication speed of xxxMbps (defined for the site, for example, Ethernet at 20Mbps). Building and automation system communication shall be via vendor's open or industry-standard protocols. The communication shall be via existing dedicated telephone lines at a speed of xxkbps. Building and controller communication shall utilize the vendor's open or industry standard protocols. The communication shall be via twisted shielded cables.

Number of pairs, wire sizing, and termination requirements shall be defined by the selected DDC vendor.

On the Facilities Network (site) level, Local Area Networks (LAN) utilize high speed communication. For example, such networks can be an Ethernet communicating over fiber optic cables at 20Mbps between the buildings, and over thin net (10baseT) wires within each building.

On the controls and automation systems level of communication, network and protocol requirements are pre-determined by DDC system vendors. However, each vendor offers several communications interfaces to adapt their systems for specific site conditions. The most suitable mode of communications for the site has to be determined in cooperation with the selected controls vendor.

On the building and controller level, the most common communication modes are over existing telephone lines, or twisted shielded pairs of dedicated wires.

Depending on the application requirements, some vendors offer other communications media such as Radio Frequency (RF), Ultrasonic, or Infrared (IR).

3. Application Software Requirements

a. Control software programs shall be loaded onto the hard disk drive of the Operator Work Station. Programs shall be downloaded to the building controllers to ensure that the control programs in the field are identical to those on record at the OWS. The program in the controller shall reside in its battery-backed-up RAM or EEPROM memory.

b. For software names, use site-specific acronyms.

c. Alarms shall be provided for analog deviations, run time totalization, range deviation, unintentional changes of state, and other relevant functions. The system shall report time of occurence, location, acronym, description of alarm, criticality, and associated alarm message. Alarms shall report to the OWS and be logged on the designated alarm printer.

Application software requirements relevant to the job should be listed in the job design documentation. The requirements in this section (Item 3) are site-specific requirements.

4. Redundancy

Sufficient redundancy shall be provided so that operations shall continue unimpaired given the failure of any single building controller or BAS communications network.

In case of a controller failure, the failed equipment will remain in the last position.

Procedures for system shutdown and recovery after failure shall be provided by the vendor.

System redundancy depends on the level of required continuous, uninterrupted operation, and the maximum tolerable off-line time for any given component of the system. More redundancy means a higher degree of reliability, at a higher cost.

For commercial (building) applications with standalone building DDC controllers, the distributed operating software provides sufficient redundancy. Each controller containing the operating software can be considered standalone. However, unitary and other controllers without "full application software" connected to the building controller will lose their software parameters and continue operating with the last parameters in case of failure of the building controller.

Industrial or laboratory applications usually require a higher degree of redundancy. To provide seamless switchover in case of failures, there may be requirements for redundant CPU, power source, communications cards, communications trunks, controllers, and I/O modules.

5. System Diagnostics

The controls and automation system shall provide the operator with sufficient information for operation and on-line troubleshooting of connected HVAC systems.

Tuning screens shall be provided to aid the operators in tuning individual PID loops from the Operator Work Stations. Historical trends shall aid the operator in diagnostics of long-term trends.

In the computer industry, software and network designers took giant steps toward developing diagnostics software. But there is a lot to be done for diagnostics of HVAC or building systems by DDC control systems. Most DDC vendors provide basic systems diagnostics (not HVAC diagnostics) with their systems. HVAC systems troubleshooting routines, defined in the DDC software, are very useful tools for the operators. Due to uniqueness of sites and HVAC systems, these have to be developed for individual applications.

6. Standard Analog Scale Ranges and System Accuracies

Facilities can decide on site-specific end-to-end tolerances and systems accuracies requested for analog values, or can make reference to existing standards.

7. Site-specific Operating Limits

Each site can develop specific operating ranges for their HVAC systems. The operating limits aid the DDC systems engineer in setting up initial operating limits, and also the operators in problem diagnostics. However, these limits can be modified for individual systems and buildings.

Items 6 and 7

Exhibit 2-5 is a table for scale range and end-to-end accuracies, reprinted from ANSI/ASHRAE Standard 114-1986. Table 2-1 is an example of site set points and their limits.

Part III—Execution of a Site Specification

The project execution section of a site specification is a description of the owner's requirements to assure structured, quality execution of the controls and automation project in all three phases: engineering, implementation, and turnover.

		Type of Usage Data				
Variable Measured	Range of Usage	Monitoring Normal	Accuracy High	On Line Control	Energy Calculation For Cost Accounting and System Optimization	Comments
Temperature:						
Space DBT	50-90°	±2°	±1°	±1°	±1/2°	All Temps. - °F
Hot Air Supply - DBT	40-140°	±5°	±2°	±2°	±1°	
Cold Air Supply - DBT	40-70°	±2°	±1°	±1°	±1°	
Outside Air DBT	−30-130°	±2°	±1°	±2°	±1°	
Dewpoint	30-80°	±3°	±3°	±3°	±3°	
Hot Water - S&R	100-250°	±5°	±2°	±2°	±2°	
Chilled Water S&R	30-100°	±2°	±1°	±1°	±1°	
Condenser Water S&R	40-120°	±5°	±2°	±2°	±2°	
Temp. Difference - Water					±0.5°	
- Air					±0.5°	Sensible Heat Cal.
Relative Humidity - Space	20-80%	±5% FS	±3% R	±3% R		FS = Full Scale
- Outside Air	20-80%	±5% FS	±3% R	±3% R		R = Range Scale
Flow - Water	Specified	±10% FS	±5% FS	±2.5% FS	±2.5% FS	FS = Full Scale
- Air	Specified	±10% FS	±5% FS	±4.0% FS	±2.5% FS	FS = Full Scale
Pressure - Air Duct	Specified	±5% FS	±2.0% FS	±2.0% FS	±1.0% FS	FS = Full Scale
- Air Building	−0.5 to +0.5" H₂O	±5% FS	±2.0% FS	±2.0% FS	±1.0% FS	FS = Full Scale
- Refrigerant Water	Specified	±5% FS	±3.0% FS	±2.0% FS	±2.0% FS	FS = Full Scale
Electric Meters				±0.25% of Reading	±0.25% of Reading	From full lag to full Lead Power Factor ANSI C12.1-82

Exhibit 2-5. Sample table showing recommended scale range and end-to-end accuracies

Systems	Operating ranges
AHU	
Free cooling	OAT = 55 – 65°F
Dampers on return air	OAT < 55°F
	OAT > 65°F
Dampers modulate on return air	OAT = 55 – 65°F
Cold deck temperature	CDT = 50 – 60°F
Hot deck temperature	HDT = 70 – 110°F
Fan discharge temperature	FDT = 50 – 80°F
Mix air temperature	MAT = 50 – 80°F
Return air temperature	RAT = 60 – 85°F
Heating	
Heating advisory prior winter	53 – 58°F (continuous for 24 hrs.)
Hot water pumps on	50°F (continuous for 4 hrs.)
Hot water pumps off prior summer	55°F
Hot water reset schedule	OAT 0°F HWST 190°F
	55°F 104°F
Cooling	
Cooling advisory prior summer	63 – 67 °F (continuous for 24 hrs.)
Chill water pumps on	60°F (continuous for 4 hrs.)
Chill water pumps off prior winter	65°F
Room temperature	
Summer	Day 72°F
	Standby 76°F
	Night uncontrolled
Winter	Day 70°F
	Standby 68°F
	Night 60°F (unless otherwise specified for controlled environment)

Table 2-1. Site Temperature Set Points and Their Limits

Part III-Execution of a Site Specification

1. Site Inspection and Engineering Coordination

The project manager shall facilitate and coordinate all activities among the owner, controls and automation vendor, and the HVAC design engineer throughout the project. All involved parties shall take the initiative to accomplish the technical and financial objectives of the project. This will include, but not be limited to, review of all relevant documentation, surveying the site and existing documentation, interviewing operations personnel, and holding mutual technical consultations.

2. Installation

a. The vendor shall locate all controls and automation instrumentation throughout the HVAC system and building. The vendor is also responsible for proper function of field instruments installed in adverse locations or conditions not recommended by the instrument manufacturer (for example, location of flow meters, installation of instruments susceptible to noise vibration, RF, transmission losses, temperature, humidity, and other ambient conditions).

b. The vendor shall supply interposing relays and other hardware not specified in the mechanical and electrical specification, but traditionally supplied by the controls contractors.

c. The controls and automation contractor shall be responsible for wire terminations at the DDC, field points, third party controllers, and communication terminals.

d. Upon completion of the installation, the controls contractor's field technicians shall completely test, verify, and tune the system, to render ready for use the complete control system.

3. Quality Assurance

The controls and automation vendor shall provide quality assurance of the installed hardware and software, and shall comply with the commissioning agent's requirements. The vendor shall submit relevant test documentation for review and approval as requested by the commissioning agent.

Item 1

Performance specifications do not have to list all conditions and requirements related to the facility, or existing systems, and do not provide detailed specifications of controls and automation system hardware and software.

It is assumed that the site performance specification will be supplemented by:

a. A detailed job specification

b. A vendor's site survey, research of existing documentation, and interviews with operating and maintenance personnel

c. Reviews and interaction with the owner's commissioning agent and HVAC engineer throughout the project.

Whether the DDC contractor is a sub-contractor to a mechanical contractor, or provides a turnkey installation, the controls and automation vendor is responsible for the installation of the system.

When wiring is installed by an electrical contractor, the controls contractor must make sure that all wires are properly terminated, labeled, and rang-through prior to testing the installed system.

Following the installation, the controls and automation contractor should complete color coding and labeling of field equipment as described in the specification. Special attention should be paid to using site-specific acronyms (synonymous with the software point definitions) to identify installed controls and automation hardware.

The project manager and the commissioning agent should schedule participation and periodical inspections of the job site by the owner's O&M personnel.

Item 3

Quality Assurance (QA) of the installed system is the most important part of the implementation process. To test all functions and components of a complex automation system controlling the connected HVAC systems requires ample time. Unfortunately, time may be short toward the end of the construction phase. This may be due to delays of installation, and pressure from the owner to meet the scheduled date for occupancy. Many unforeseen events can cut into the time slot allocated for systems testing. Frequent systems "glitches" during early stages of building occupancy point to inadequate QA and toward the need for more QA during installation and start-up. Ongoing QA during design and implementation phases will reduce the number of field QA problems and will assure compliance with technical objectives and parameters defined in the design documentation.

4. Field Tests

The owner's commissioning agent may provide the controls vendor with a list of required tests. In the absence of such a document, the following can serve as a guideline for testing:

  a. Complete field tests shall be performed on all sub-systems.
  b. Each individual function shall be tested, and proven correct in function and response, a minimum of two times—during testing and test operation.
  c. Verification testing for the owner shall occur within, and no later than, two months after the vendor's tests. The controls and automation contractor shall provide a fully qualified service engineer who, together with the owner's authorized representative, shall perform the verification tests. Owner verification tests should be scheduled and performed after the controls contractor is satisfied that the automation system's field gear, software, and communications are adjusted and operating in accordance with specification requirements.

5. Test Procedures

a. Control Panels

Controls hardware testing and QA shall be conducted by the vendor prior to shipment of the panels to the job site. On-site testing, component calibration, and jumper and switch settings of the installed hardware shall be completed prior to testing of control loops. Test protocols shall be submitted to the owner's commissioning agent as part of the turn-over documentation.

b. Control Function Testing

1. All output channels, both analog and digital, shall be commanded to their respective positions, such as on/off, stop/start, or adjusted to a certain value, and their operation verified at the field device.
2. All analog input channels shall be verified for proper operation and accuracy. The accuracy should be demonstrated by (a) calibration sheets, and (b) by comparing the DDC system readings with an external calibration reference standard.
3. All digital input channels shall be verified by changing the status of the field device, and observing the appropriate displayed change of state.
4. If a point should fail testing, the controls contractor shall perform necessary repair action and re-test failed points and all other interlocked and software points associated with the repaired point.
5. Automatic control operation of a control loop or a series of control loops cascaded, shall be verified by introducing an "error" into the system and observing the proper corrective system response. The operation shall be verified at the field device, and at the OWS.
6. Selected time schedule programs and set points shall be verified by changing the set variables in the programs and observing the correct response on the controlled output as verified at the OWS terminal and field device.
7. The contractor shall demonstrate tuning parameters, and verify the response of the actuators for each control (PID) loop.
8. The contractor shall demonstrate the test results of all communication protocols to the owner. The contractor shall also demonstrate communication at all levels of the vendor's architecture, including communication to connected third party controllers. This shall be done by verifying the function of each point (field or software point), or displayed values of points mapped through the communication protocol. The transmission verification shall be demonstrated at all nodes of the system, with emphasis on the data transmission to the designated controllers and OWS on the network.

c. Operator Work Station (OWS) Test Procedures

1. Communication with each DDC controller shall be demonstrated.
2. Operator commands shall be explained.
3. PMI compliance with as-built documentation, operating manuals, and the existing field installation shall be demonstrated. Colorgraphics shall include all components of the depicted HVAC equipment, status, measured and set values for each installed point, and mode of operation. Each graphic screen shall include instructions for the operators to help them move from one screen to another, or to the main menu. Colorgraphic screens shall be dynamically updated at a xx second scan rate, as published in the vendor's documentation.
4. Communication verification to each field point shall be demonstrated by the owner's commissioning agent randomly selecting and accessing points in each controller.

5. Operator commands, system shutdown/re-boot, and software and data base maintenance procedures shall be demonstrated.

6. Access of field data and DDC files via the OWS, data base download/upload to the controllers, read/write to existing files, custom programming, file copy, and other software features shall be demonstrated.

7. Time outs, retries, and system and communication error checks shall be demonstrated.

8. Transmit value/change over the network shall be demonstrated.

9. Avoidance of nuisance alarms, inhibition of alarms for devices out of service, HVAC system shutdown/start-up, and other HVAC controls related features shall be demonstrated.

10. Messages and advisories related to operation and safety shall be demonstrated.

11. Report compilation, ad hoc report generation, system data printout, history and trend data printout formatting, and how to transfer to other media available on the network shall be demonstrated.

12. Control loop functions and PID loop tuning from the OWS shall be demonstrated.

13. Optimization and energy management functions shall be demonstrated by changing input parameters to simulate real life situations.

d. Performance Verification Testing

The systems performance shall be verified by its proper operation, including, but not limited to, the following:
- Continuous trending of key parameters (temperature, humidity, pressure, on/off time, etc.)
- Alarm logs
- Error logs
- Trending of PID loop performances
- Change of parameters
- Controllers standalone operation
- Software upload/download, etc.

After the 30 day performance test and acceptance of the system by the owner, the warranty period, as described in the Warranty requirements, shall begin.

Training

Building HVAC as well as DDC control and automation systems are getting more and more complex, and only properly trained staff can assure their optimum operation. Building owners should establish continuous training programs for the system operators and maintenance personnel, to assure their proficiency.

If the site already has an existing controls and automation system, and the same or a similar system is added during the site renovation, the requirements for DDC training may be waived. However, training on building-specific operations should be part of the job contract.

For sites with new DDC systems, training requirements should be part of the job specification. All major vendors offer formal training classes for systems engineering, and operation and maintenance. Training requirements should be included in the DDC vendor's job contract.

BAS Material Specification

Even though performance specifications call for fully operational BAS systems and put the burden for systems performance on the selected BAS vendor, standardization of selected hardware for the facility is prudent.

BAS material specifications set standards for the hardware and field equipment most appropriate for the facility from the operations and maintenance standpoint. Standardization of selected DDC systems and hardware components economizes on spare parts inventory, material management, and training of controls technicians and controls maintenance and repair. The material specification should be published in every job documentation related to the site. Exceptions from the specified material or equipment should be granted by the owner's plant engineering or O&M department only after careful review of the proposed substitutions.

Field Hardware Specification

Hardware specification in this section describes the components most commonly used for HVAC controls and automation systems. Because it is site specific, and aimed at long-range facilities development and standardization, the owner should carefully screen the hardware available on the market. Service technicians appreciate hardware and components that require minimum maintenance, have a minimum failure rate, and are easy to use and maintain. Service technicians should be invited to contribute to the development of the material specifications.

Even with the most carefully selected hardware, there will be replacements of models and parts over time. Depending on the site, the specification should define equivalent or comparable components by manufacturer and model. "Approved equals" should be included in the site specification, in a table format for easy reference. Spare parts can then be stocked on the basis of market prices and failure rates.

Some sites may require competitive pricing for purchases over a certain dollar value. In such cases, the site can define only the main characteristics of individual hardware. Selection of make and models would then be defined by the specifying engineer for new jobs, or material management for spare parts. In either case, the hardware specification should be reviewed with:

    a. The owner's plant engineering department to assure compliance with standardization goals of the facility

    b. The material management department to assure favorable pricing and maintenance of minimum stock levels

    c. The O&M department to assure required tooling, instrumentation, and training for long-term hardware maintenance.

The following specification describes the most commonly used HVAC controls and automation hardware components.

Remote DDC Controllers

    a. Each DDC controller shall be enclosed in a NEMA 1 enclosure or cabinet. The controller shall be configured so that the enclosure can be mounted, and electrical terminations made prior to the installation of the control electronics. The controller cabinet shall be populated with

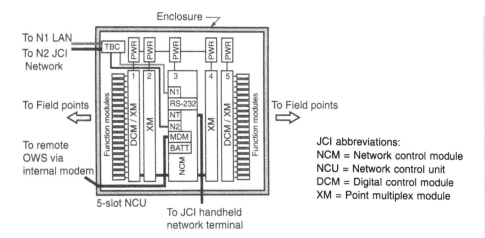

Figure 2-6. Example of a Johnson Controls, Inc., DDC controller

JCI abbreviations:
NCM = Network control module
NCU = Network control unit
DCM = Digital control module
XM = Point multiplex module

the modular DDC electronics during the start up/check out phase of the project. All control cabinets shall be outfitted with a master keyed lock, and shall have a 15 Amp 120VAC duplex outlet.

b. The DDC controller shall contain all processors, point modules, and interface modules, for controlling the connected field equipment and communicating on the associated network.

c. Each DDC controller shall have 10% spare capacity. For DDC controllers designed for specific applications provide a spare controller of each type utilized in the project.

d. Control, communication, and power circuits shall be individually electronically isolated to protect against transients, spikes, and power surges.

e. Each DDC controller shall save its programming in non-volatile memory (EPROM) or have a battery backup with internal charger to provide necessary backup power for RAM, lasting up to 72 hours.

f. The DDC controller shall be UL approved, with a UL listing as a Signaling System.

Field Instruments

Dampers and Damper Actuators

a. Automatic dampers shall be of modular sections. Maximum size of any one section is less than 48″ x 48″. The frame shall be constructed of 13-gauge galvanized steel. Blades shall be double 22-gauge galvanized steel with square or hex drive pins of a minimum 1/2″ diameter. Bearings are oil-impregnated sintered bronze. Side seals shall be spring loaded stainless steel. Blade seals shall be silicone or Santoprene. Two-position dampers shall be parallel blade; modulating dampers are opposed blade. Leakage shall not exceed 3 CFM per square foot at 1″ W.G.

Approved dampers for the site are:
(a) Make and model for opposed blade type
(b) Make and model for parallel blade type
Other types must be approved by the Plant Engineering department.

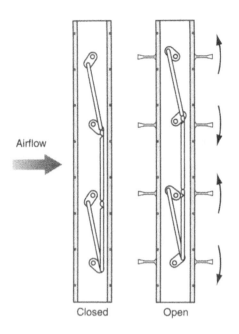

Airflow →

Closed   Open

Figure 2-7. Automatic damper with opposed blades

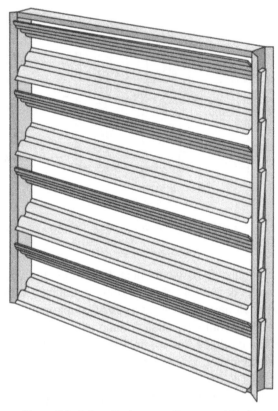

Figure 2-8. Automatic damper with opposed blades

b. Smoke dampers shall be of modular sections. Bearings shall be self-lubricating porous bronze. Side seals shall be spring loaded stainless steel. Blade seals shall be rated to withstand 250°F.

Smoke dampers shall be UL 555S listed and bear the UL label. Leakage rating under UL 555S is leakage Class II (10 CFM per square foot at 1″ W.G.). Appropriate electric actuators will meet all applicable UL 555S qualifications for

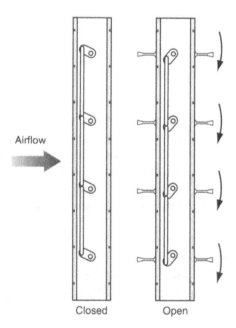

Airflow →

Closed   Open

Figure 2-9. Automatic damper with parallel blades

FireStat

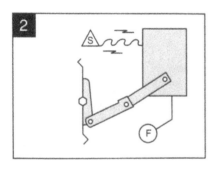

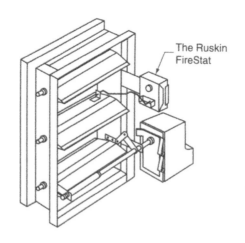

The Ruskin
FireStat

1. Damper is equipped with electric FireStat instead of fusible link. Actuator maintains damper in open position.

2. Smoke detector or FireStat can close damper. After response to smoke or fire, the damper can be tested or reset from a remote station.

3. When damper is closed in response to smoke or fire, firefighting commander can override FireStat and/or smoke detector to operate damper as part of dynamic smoke management system. Damper operation capabilities are maintained regardless of the presence of fire conditions or smoke.

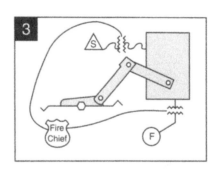

Figure 2-10. The Ruskin FireStat damper control

both dampers and actuators. Smoke dampers are to be supplied with the motor factory mounted, and with an end-switch to signal open position.

Approved smoke dampers are:

(a) Make and model

(b) Substitute make and model

Other types must be approved by the Plant Engineering department.

c. Exhaust hood dampers (Figures 2-11 and 2-12) shall be of modular sections. Bearings shall be stainless steel sleeve type. Side seals shall be spring loaded stainless steel.

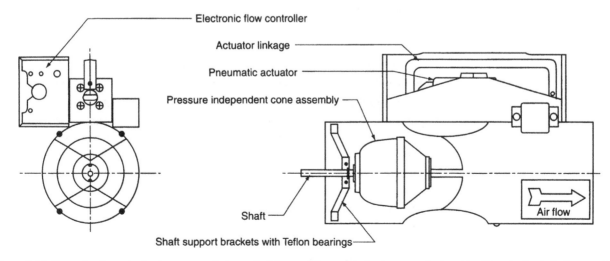

Figure 2-11. Example of an electronic pressure independent linear airflow control valve, manufactured by Phoenix Controls Corporation

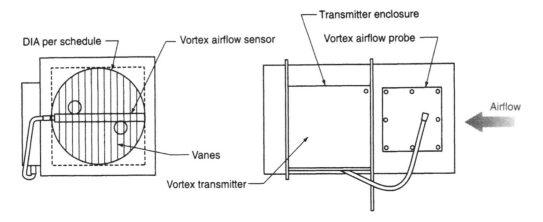

Figure 2-12. Example of pneumavalve with vortex air flow sensor, manufactured by TEK-AIR Systems, Inc.

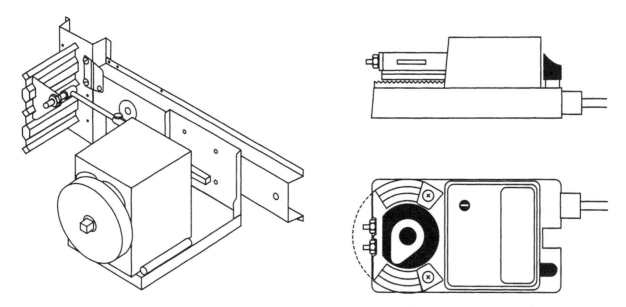

Figure 2-13. Damper actuator with linkage spring return

Figure 2-14. Electric direct coupled damper actuator

Approved dampers are:
(a) Make and model
(b) Substitute make and model
Other types must be approved by the Plant Engineering department.
Approved damper actuators for the site shall be electrical actuators.

d. Electrical damper actuators (Figures 2-13 and 2-14) shall be provided with a spring return. Actuators shall be sized for 125% of the required torque of the controlled dampers. All actuators shall include a feedback mechanism that senses the actual position and provides positive feedback to the DDC controller. Feedback taken from controller command output shall not be acceptable. All actuators shall be labeled with the software name.

Approved damper actuators for the site are:
(a) Electrical actuators—make and model
(b) Substitute electrical actuators—make and model
Other types must be approved by the Plant Engineering department.

The owner has to decide whether to use electrical (Figures 2-13 and 2-14) or pneumatic actuators (Figure 2-15). For years, there has been an ongoing discussion on electric versus pneumatic actuators for DDC control systems. Popular belief is that pneumatic actuators are more reliable and less expensive. Such perceptions have two roots:

1. Simplicity of pneumatic actuators

2. Long tradition in using pneumatic actuators.

If the majority of controls in the building are electronic, the limited number of pneumatic actuators would require installation and maintenance of a pneumatic system including air compressor, oil separator, air dryer, and a distribution network of pneumatic tubing. The reliability of electrical actuators has increased markedly, and their failure rate is minimal. There are however many existing facilities with operating pneumatic installations, and sites with preferences for pneumatic actuators.

2. Automatic Control Valves

All automatic control valves required by the control system shall be provided by the controls and automation contractor for installation by the Division 15000 Contractor. The valves' fail-safe position shall be normally open or closed, as defined by the HVAC engineer. Modulating, single seated, straight through valves for chilled water, hot water, or steam service shall be provided with equal percentage contoured throttling plugs.

Valves 2″ and smaller shall be ANSI class 250, brass body, or cast iron screwed connections. Valves larger than 2″ shall be ANSI class 125, with cast iron body and flange connections. Stems shall be stainless steel with "spring-ring" type packing guaranteed against leakage for one year past contractual warranty, with only a packing nut adjustment required. Maximum allowed pressure drop across the control valve shall be 5 psi.

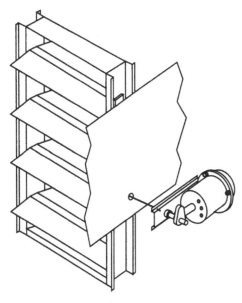

Figure 2-15. Pneumatic actuator

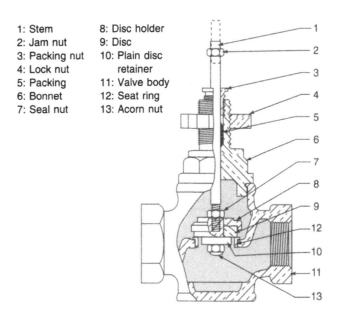

1: Stem	8: Disc holder
2: Jam nut	9: Disc
3: Packing nut	10: Plain disc
4: Lock nut	retainer
5: Packing	11: Valve body
6: Bonnet	12: Seat ring
7: Seal nut	13: Acorn nut

Figure 2-16. Two-way globe control valve for steam and water

Approved valves are:
(a) Make and model
(b) Substitute make and model
Other types must be approved by the Plant Engineering department.

The valve schedule in the design documentation must indicate capacity, pressure drop, size, and Cv rating. All installed control valves should operate smoothly without hunting or cycling, and should be capable of operating within the temperature and pressure range of the controlled fluid without leakage through the packing.

3. Electrical Valve Actuators

a. Electrical valve actuators shall be provided with a spring return. Actuators shall be sized for 125% of the required torque of the controlled valves. All actuators shall include a feedback mechanism that senses the actual position and provides a positive feedback to the DDC controller. Feedback taken from the controller command output shall not be acceptable. All actuators shall be labeled with a software name.

Approved electrical valve actuators for the site are:
(a) Electrical actuators—make and model
(b) Substitute electrical actuators—make and model
Other types must be approved by the Plant Engineering department.

Valve actuators are either pneumatic or electrical (Figure 2-17). This specification is an example of an electrical valve actuator. However, many facilities have and/or install pneumatic actuators for new or retrofit jobs (Figure 2-18).

Positive feedback from the valve position is desirable for systems troubleshooting by the operators or maintenance personnel. Where the sensors are close to the controlled device, for example, a coil, the sensor's analog input can substitute for the feedback.

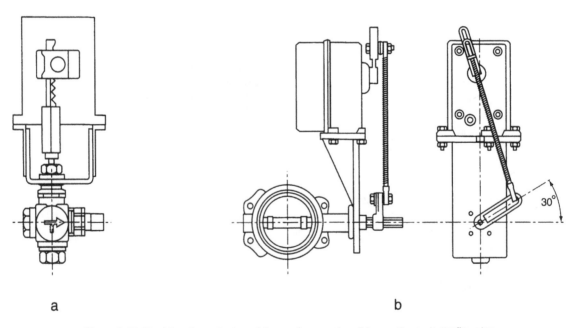

a   b

Figure 2-17. Electric valve actuators: (a) on a 3-way valve; (b) on a 2-way butterfly valve

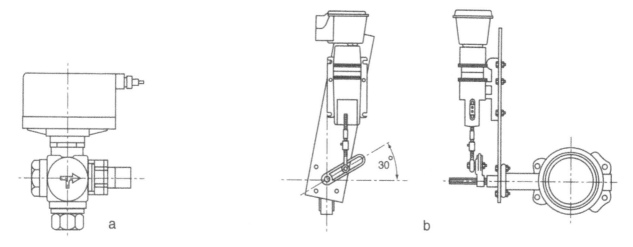

Figure 2-18. Pneumatic valve actuators: (a) on a 3-way valve; (b) on a 2-way butterfly valve

4. Thermostats

Not specified in this specification.

The use of thermostats (T-stats) for specific applications must be approved by the Plant Engineering department and specified under Division 15000 or 16000 of the job specification.

5. Low Temperature Detection

a. Low temperature detection sensors (freeze stats) shall be snap acting with manual reset. One set of normally closed (N.C.) contacts shall be hardwired to the motor starter of the fan; the second set of contacts is utilized for alarm input to the DDC controller. The freeze stats capillary shall have a minimum sensitive length of 20 feet, and shall be installed in a serpentine fashion a maximum of 18 inches downstream of the coil it is protecting. Freeze stat activation occurs if any two (2) inch section of capillary is below set point.

   Installation of one freeze stat per 20 square feet of coil to be protected is required. Multiple freeze stats must be electrically wired in series.

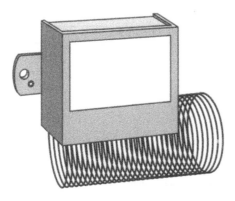

Figure 2-19. Electric low limit sensor

Approved low temperature detection devices are:

(a) Make and model

(b) Substitute make and model

Other types must be approved by the Plant Engineering department.

6. Low Steam Pressure

a. Low steam pressure switches (Figure 2-20) shall be in the 0 to 15 PSIG range, with contacts opening on pressure decrease, and snap acting with automatic reset.

Switches are to be supplied with the siphon "pig tail."

Approved low steam pressure switches are:

(a) Make and model

(b) Substitute make and model

Other types must be approved by the Plant Engineering department.

There are several types of temperature sensors applicable to any DDC system. The facility should consult with the systems vendors on acceptable types of temperature elements for the particular systems used at the site.

The following specification relates to resistance temperature devices (RTD sensors).

---

7. Temperature Sensors

Temperature sensors shall be resistance temperature devices (RTDs) having an accuracy of not less than 1% across full scale range. Sensors shall be platinum or nickel wound with a reference resistance of 1000 ohms at 70°F.

a. Space temperature sensors shall be concealed by brushed nickel covers secured with tamper proof screws.

Sensor accuracy shall be better that 1%.

Approved space temperature sensors are:

(a) Make and model

(b) Substitute make and model

Other types must be approved by the Plant Engineering department.

b. Duct temperature sensors shall be single point type, except for mixed air and discharge air applications where an averaging type element is to be used.

Sensor accuracy shall be better than 1%.

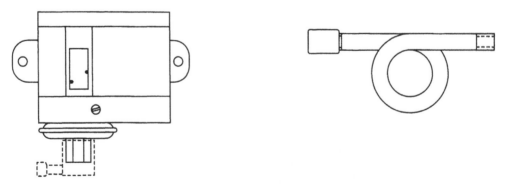

Figure 2-20. Low pressure switch

Approved duct temperature sensors are:

(a) Make and model

(b) Substitute make and model

Other types must be approved by the Plant Engineering department.

c. Well insertion temperature sensors (Figure 2-21) shall be provided with brass or stainless steel wells, and the assembly rated for the temperatures and pressures of the locations where installed. Condensate temperature sensors are to be provided with stainless steel wells, and rated for operating temperatures up to 550°F.

Sensor accuracy shall be better than 1%.

Approved insertion temperature sensors are:

(a) Make and model

(b) Substitute make and model

Other types must be approved by the Plant Engineering department.

There are several types of humidity sensors used with DDC applications. The owner should consult with the systems vendors on acceptable types of humidity elements for the particular DDC system installed at the site. The following specification relates to capacitance type sensors (Figure 2-22).

---

8. Humidity Sensors

Humidity sensors shall incorporate capacitance type sensing with a 3% accuracy over a range of 20-80% RH (relative humidity).

a. Space humidity sensors shall be concealed by brushed nickel covers secured with tamper proof screws.

Approved space humidity sensors are:

(a) Make and model

(b) Substitute make and model

Other types must be approved by the Plant Engineering department.

b. Duct humidity sensors shall be capable of withstanding temperatures from −40°F to 120°F, and humidity from 10 to 100% RH without damage or need for re-calibration.

Approved duct humidity sensors are:

(a) Make and model

(b) Substitute make and model

Other types must be approved by the Plant Engineering department.

9. Differential Pressure Transmitters (Air/Air)

Differential pressure transmitters shall be variable capacitance type with a stainless steel diaphragm in a stainless steel body. Excitation voltage shall be 24VDC with reverse polarity protection. Transmitter output shall be 4-20MA.

Pressure ranges shall be selected to match the operating range of systems. Units shall be provided with an over-pressure protection of five (5) times full scale, 10 inch W.C. minimum. Accuracy is 1% of full scale.

Approved differential pressure transmitters are:

(a) Make and model

(b) Substitute make and model

Other types must be approved by the Plant Engineering department.

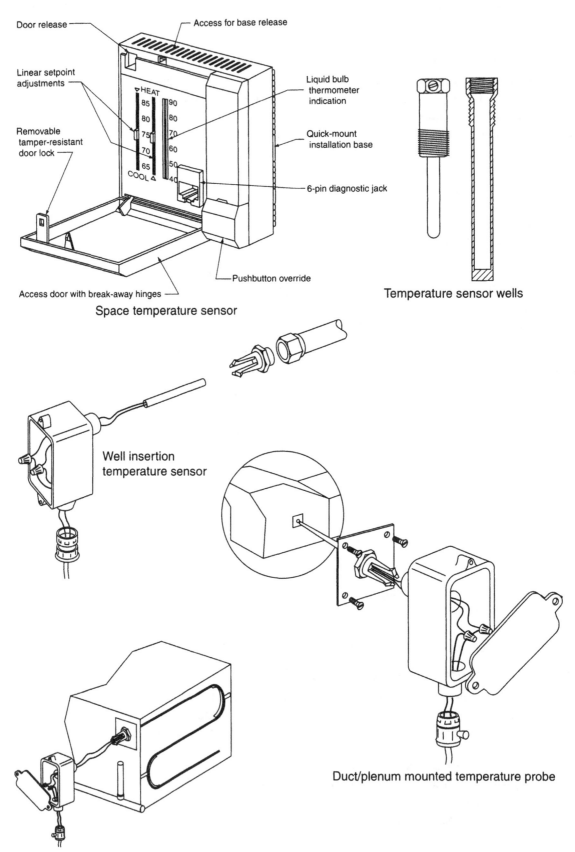

Door release — **Access for base release**

Linear setpoint adjustments

Removable tamper-resistant door lock

Liquid bulb thermometer indication

Quick-mount installation base

6-pin diagnostic jack

HEAT
85 — 90
80 — 80
75 — 70
70 — 60
65 — 50
COOL — 40

Access door with break-away hinges

Pushbutton override

Space temperature sensor

Temperature sensor wells

Well insertion temperature sensor

Duct/plenum mounted temperature probe

Temperature duct averaging sensor

Figure 2-21. Temperature sensors

64

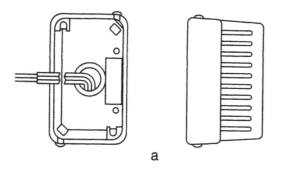

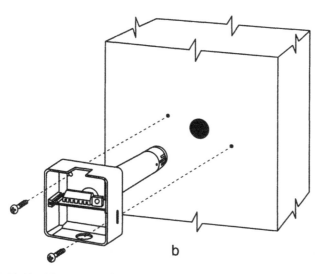

Figure 2-22. Humidity sensors: (a) space humidity sensor; (b) direct humidity sensor

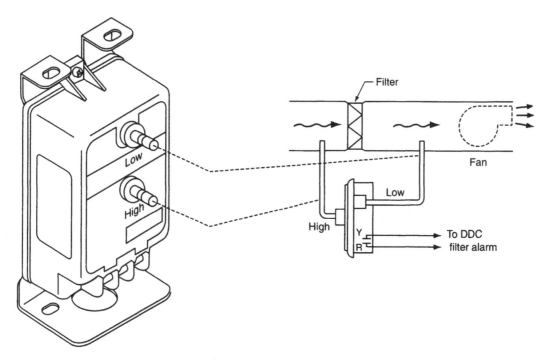

Figure 2-23. Air-air differential pressure transmitter

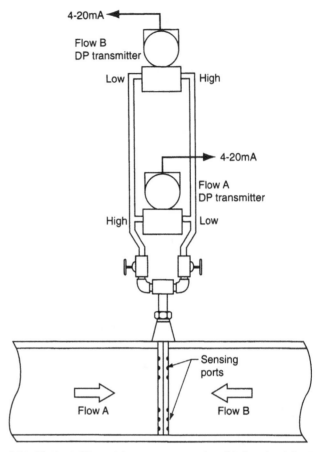

4-20mA

Flow B
DP transmitter

Low   High

4-20mA

Flow A
DP transmitter

High   Low

Sensing
ports

Flow A   Flow B

Figure 2-24. Wet/wet differential pressure transmitter (bi-directional flow)

10. Differential Pressure Transmitters (Wet/Wet)

Differential pressure transmitters shall be variable capacitance type with a stainless steel diaphragm in a stainless steel body. Excitation voltage shall be 24VDC with reverse polarity protection. Transmitter output shall be 4-20mA.

Pressure ranges shall be selected to match an operating range of systems. Units shall be provided with over-pressure protection of ten (10) times full scale. Accuracy shall be 0.25% of full scale. Transmitters shall be provided with a three-valve manifold constructed of materials matching the installed piping.

Approved differential pressure transmitters are:

(a) Make and model

(b) Substitute make and model

Other types must be approved by the Plant Engineering department.

11. Control Devices and Transducers

Output relays for start/stop control shall be plug-in type, 3PDT 10A rated. Relays shall be mounted at, or within, the motor starters, or in the field controls equipment cabinet.

12. Variable Frequency Drives

Variable Frequency Drive (VFD) shall be a solid state variable frequency controller in a NEMA 1 enclosure. The controller shall be UL listed. The controller shall control the motor's AC power in a 10% to 110% range, in a stepless mode. The VFD shall be designed for the following ambient conditions: elevation to

3300 feet without derating, temperature 0°F - 45°F, relative humidity to 95% non-condensing, AC voltage variation −5 to +10%, five-cycle regulator control power dip-through.

The VFD shall have the following features: start/stop, speed selection, Hz controlled speed range, auxiliary contact for alarm indication, current limit, automatic restart, one plug-in test cord for troubleshooting, and analog input/output signals for 4-20 mA.

The VFD shall include: AC input disconnect, motor thermal overload protection, and manual by-pass of the VFD by a "controller-off-by-pass" push-to-turn transfer switch.

The VFD manufacturer shall have at least ten years experience in manufacturing VFD controllers.

Approved differential pressure transmitters are:

(a) Make and model

(b) Substitute make and model

Other types must be approved by the Plant Engineering department.

Items 11 and 12

In the past, variable frequency drives were typically specified and installed by Division 15000 mechanical contractors, and wired by the job electrical contractor. (Traditionally, all hardware related to the air handling unit, such as fan, motor, and inlet vanes, were part of the mechanical trade, and supplied by the mechanical contractor).

With the advancements of automation, VFD features, and their inclusion in controls strategy (air volumetric control), the VFD could be included in the controls specification.

Spare Parts

A proposed Spare Parts List should be developed in cooperation with the DDC vendor, and included in the bid. This list should include field instruments, DDC components, third party purchased components, and other hardware components the owner should have in stock.

Chapter 3

Program Development Phase

Global Approach to Facilities Automation Development

The design and implementation of environmental control systems is a complex undertaking because it combines several engineering disciplines under one umbrella. Each of the disciplines is complex on its own—combining them can be overwhelming for many engineers and project managers; therefore, teamwork among several engineering organizations and specialists is essential for a successful project.

The organizations most involved in building and systems design are architectural and engineering firms. Typically, participation by consultants that specialize in Building Automation Systems design (i.e., DDC) is limited. Part of the reason is that the design experience of building automation consultants is limited to certain DDC vendors' products. The void in consulting services is filled by the DDC vendors, who provide systems engineering support for their systems. However, this support is available only at later phases of the project, many times after the vendor has been selected.

This traditional structure of design teams leaves the owner with a gap—without systems engineering expertise in the most important phases of conceptualization and design of buildings and building systems. The gap is narrowed later by participation of the selected DDC vendor in the process, but only after design completion, bidding, and final selection of a DDC system. By then, all major characteristics of the building and its operation are determined, designed, and budgeted.

The Programmatic Approach

During program development of the project, the design team defines the main characteristics of the project, such as architectural impact of the buildings on the neighborhood, building architecture (exterior, interior), exits, roads, and electrical, mechanical and other systems. Because systems engineers (systems integrators or automation experts) are seldom on the program development team, issues related to facilities automation are not defined. Facilities automation is a term used for real time automation systems (such as BAS or DDC systems), industrial automation systems (such as power plant automation), and utilities metering, as well as data base systems such as office automation (e.g., work groups for Windows, accounting systems), inventory control systems, and maintenance management systems.

The design team should develop a global approach to program definition, and should include a definition of facilities automation systems, and their main characteristics and networking requirements in the program definition. This should also include a definition of all real time and database systems, and requirements for their interoperability. An example is a definition of DDC systems, their architecture, networking, and interface to maintenance management. For existing facilities, the definition should include integration of new systems into existing systems and networks.

Most design teams lack the expertise necessary for a programmatic approach to design of facilities automation systems and their networking, a task so important for modern facilities. The global approach to facilities automation (defining all automation needs now so that it's an integral part of the building design) is left out of most program development and design phases, leaving the owner in a disadvantageous situation. The negative consequences become apparent later, when the owner decides to integrate existing automation systems. Integration of incompatible Building Automation Systems, installed by various DDC vendors over time, is always a challenge; in many cases it may be prohibitive due to its high cost.

Lack of a unified approach to facilities automation becomes more pronounced in multi-building applications that have ongoing construction activities. The consequences can be astonishing because each building is designed or renovated at different times, by different design teams employing different design philosophies and systems selection procedures. Without a global approach to automation, in particular to HVAC controls and automation, and always selecting the lowest DDC bidders, the facility often ends up with several incompatible control systems. Having several low cost, incompatible systems on site results in the following problems for the owner:

- Proprietary controls and DDC systems are unable to communicate with one another or with a common front end, resulting in multiple "discrete isles" of automation.

- Each system requires separate hardware inventory, training of O&M personnel, and administration of multiple service contracts.

- Due to their low cost, the systems provide (in most cases) only basic control and monitoring functions.

Thus the lack of a global approach to facilities automation (and the environmental control systems are a significant part of facilities automation) results in inadequate controls, insufficient level of automation, and increased operating and maintenance costs for the owner.

To be successful in integrating facilities automation into the design/development process, follow these three approaches:

1. Establish a design group that includes BAS systems and facility engineers with a true teamwork spirit.

2. Standardize systems performance, operation and maintenance requirements, DDC systems, controls hardware, operating logic, systems documentation, networks, and so on.

3. Define quality requirements and measures, and use a structured approach to the quality review process (Commissioning/Total Quality Management approach) throughout all phases of design/development.

Implementation of any of the three approaches during the design and implementation process will lead to higher quality systems and a lower life cycle cost of installed DDC systems. The combination of the three approaches represents a global approach to DDC systems implementation with long-term benefits to the owner.

Role of the Commissioning Agent in the Project Phases

Project managers focus on the management of design teams, and on the coordination and administration of related activities throughout all phases of the project. An appointed commissioning agent should aid the project manager to focus on quality by developing and implementing a structured review process throughout all phases of the project.

A structured approach to quality systems design and implementation should add value to the traditional design and implementation process. The commissioning agent, in cooperation with project management, should develop specific goals and quality checks for individual project phases. This shared responsibility between the project manager and the commissioning agent enhances the project by focusing on quality. The project manager can devote his or her full attention to managing and administering the process, relying on the commissioning agent to provide quality assurance for design, installation, and systems performance.

The review tasks (from the Commissioning/Total Quality Management approach) should match the traditional design and implementation phases of the project (see Fig. 3-1). The commissioning agent and the review team evaluate every phase of the project prior to moving to the next phase. Though shortcuts are the way of life in the competitive engineering and construction environment, they can create unfavorable situations for controls and automation because of the systems' complexity and the numerous conditions which influence the controlled environment. Shortcuts taken during the design and engineering phase of a project can eventually turn into a nightmare during systems start-up and operation.

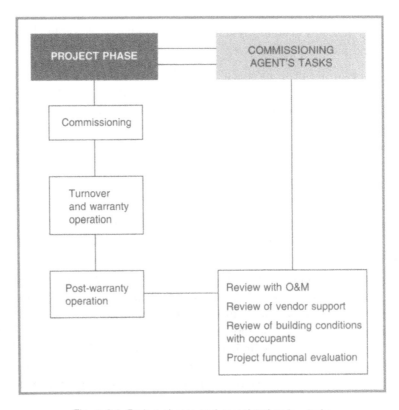

Figure 3-1. Project phases and associated review tasks

Figure 3-1 shows some of the major commissioning/TQM tasks and their interrelationships with individual project phases. Since the focus is on the commissioning agent's tasks, project management tasks are not listed.

Establishing Realistic Design Criteria in the Program Development Phase

Establishing realistic design criteria at the early stages of project or program definition is essential. For new construction or major renovation projects, the design team and the project management have to balance the owner's requirements and expectations with design constraints, implementation schedules, and overall budget.

Even though these concerns seem to be far removed from the environmental controls systems, in fact, many future problems and disappointments with the building's operation start at the conceptual phase of the project. Unrealistic design proposals and concepts can boost the owner's expectations. These are expectations which the design team cannot meet fully under the constraints of the project, particularly the expectations related to environmental control systems operation which have been created without the participation of a controls and automation systems expert.

The Program Development phase of the project should always include engineers with systems expertise. Presence of a systems engineer on the team may influence the architectural layout, space utilization, and zoning, and may provide productive ideas for HVAC design solutions. Such projects provide a higher degree of automation by fully using the applicable features and capabilities of the DDC system, and provide systems interfaces according to the overall goals for facilities automation. On the other hand, absence of such expertise from the design team can result in inadequate control of building environment and future needs for systems modifications. This in turn results in additional expenditure, a dissatisfied owner, and dissatisfied building occupants.

Definition of Areas Important for Environmental System Performance

Definition of the future building and its operating characteristics are being determined at this early phase of the project. But these definitions should also include consideration of overall facilities automation, and in particular, definition of characteristics of the future environmental control systems, including the BAS. To have a good environmental control system definition, the HVAC engineering expert should be complemented by two additional resources:

1. BAS systems engineering expertise, and

2. HVAC and BAS operations and maintenance expertise.

At this phase of the project, definition of four (actually, five) major areas will impact the performance of the future environmental controls system:

1. The building envelope, including all large openings, such as windows, doors, loading docks, etc.

2. Building layout, zoning, space utilization, etc.

3. Building operating parameters, such as room temperatures, relative humidities, air exchange rates, pressures, etc.

4. HVAC systems, their control, capacities, interfaces, coordination with site utilities and adjacent areas, etc.

5. For modern buildings, add the definition of automation and networking requirements for the facility; definition of Building Automation Systems, including DDC systems; level of desired automation; and site-specific O&M requirements.

These definitions should include the participation of systems engineers and owner's facilities engineers (or O&M personnel) with environmental controls (HVAC and DDC systems) operating experience. The definitions are important for continuous control of the building parameters by off-the-shelf DDC systems and DDC system interfaces, and include communications to other facilities.

Again, using engineers with controls experience for design definitions and reviews could avoid future systems performance and maintenance problems. Their operating experience used as feedback in the design process could prevent costly mistakes.

Case Studies

The following case illustrates the practicality of early review and feedback.

Case 1

During program definition meetings the architects and the project managers have agreed to provide operable windows for the newly designed office building. The architects' preference is based on the esthetics of the building; the project manager believes that operable windows would be appreciated by the occupants. The HVAC engineer proposes a Variable Air Volume (VAV) system with VAV boxes in the perimeter offices.

At a design review, an engineer with systems operating experience reviews the documentation and comments on inherent problems associated with VAV applications in buildings with operable windows. (Opening windows drastically changes room temperatures. Depending on the season, the VAV controller compensates for the change by opening or closing the VAV dampers which consequently change duct pressure and fan speed. The change can have an impact on all other rooms connected to the same VAV system).

Following the review, the project team decides to modify the design to fixed windows. Notice how the early review provided input at a time when design modifications could be easily accomplished and have no impact on the project cost. A simple modification at an early stage of the project could later have become a major problem for control of the desired building parameters.

Meeting the owner's specifications is one of the most important tasks of any design team. However, in some cases the owner's requirements may be unrealistic, beyond the technical feasibility of commercially available systems, or over the available budget. The design team should provide the owner with a realistic scope of work, and define a project which provides the specified (and expected) systems performance. The following case illustrates contradictory conditions and their possible solution.

Case 2

An existing library planned renovation of its HVAC systems, and an upgrade of associated controls to DDC. The library was a stone building with single-pane stained glass windows. All library areas, including the book stacks, were open to the public. To preserve the books, the curators requested tight temperature and humidity controls for the entire library. The designated temperature was to be controlled within $+/- 1°F$, and humidity within $+/- 3\%$ for year-round (summer/winter) operation. As with most projects, the budget was limited.

The design team (with controls experience from operating similar buildings) surveyed the existing situation and found the following:

- Uninsulated stone walls which allowed moisture to penetrate
- Single pane windows which allowed air and moisture to penetrate, and collected condensation
- Book stacks in public access areas

The team recommended (a) insulating the exterior walls, including moisture barriers, (b) modifying the windows to double-pane windows, and (c) separating the book stack area from the public access areas. Separation of book stacks from public access and office space allowed control of the book stacks in required tolerances, while the rest of the building could be controlled as an office building. Without these recommendations it would have been difficult to maintain the required parameters in all library areas year round. While oversized and over-designed HVAC and control systems could maintain the required parameters under favorable ambient conditions, the cost of implementation, energy, and maintenance of such systems would be excessive.

Without visualization of the environmental control systems performance at the project definition phase by engineers with DDC systems operating experience, these projects would have resulted in erratic building pressures, temperatures, and humidity conditions. These would have been blamed on the environmental control systems, more specifically on the DDC control systems, and both cases would have become AIA statistics in the category of complaints related to occupancy comfort.

Design problems in these cases would not show up immediately upon completion of the job and full occupancy, especially if the turn-over took place during mild spring ambient conditions. They would become apparent later on, with the seasonal changes and arrival of hot and humid summer, or on cold, rainy winter days. It would have taken many site visits and surveys to discover the true origins of the problem.

These cases are not fairy tales. Similar scenarios go undetected all the time. Problems similar to these are not, and should not, be classified as environmental controls problems. Their origin is in the building design. They manifest themselves as environmental controls-related because (a) the HVAC and DDC systems are unable to maintain the required parameters in the designed building, and (b) the problems are detected and reported via the DDC system. Design deficiencies become chronic problems, resulting in excessive O&M costs for systems maintenance.

Without the participation of engineers who have real life controls and operating experience, in the design team or in a structured review process, there is always a possibility of not fully recognizing future operational problems. This is especially true at this early design phase of conceptualization and visualization of the behavior of the future building.

Cost of Modifications During the Life of the Project

Structuring project teams for performance design, and implementing a structured review process, will lead to a building or systems design that will meet more fully the required operating parameters.

TQM procedures in manufacturing were implemented for early detection of potential problems. It is considered more cost effective to implement TQM at each stage of the process, as opposed to paying for repairs, changes, and modifications at later stages, or paying even more by losing customers due to the low quality of manufactured products or services.

A representation of the escalation of costs associated with changes and modifications at individual stages of design and implementation process is shown in Figure 3-2.

Developing the Commissioning/TQM Process

At the beginning of the program development phase, the commissioning agent should establish three basic TQM measures:

1. Define long-term objectives and assembly of teams
2. Establish performance requirements and measures
3. Identify customers for every task

Project Objectives

For definition of long-term objectives, the commissioning agent has to look beyond the design and even beyond the installation process to set global ob-

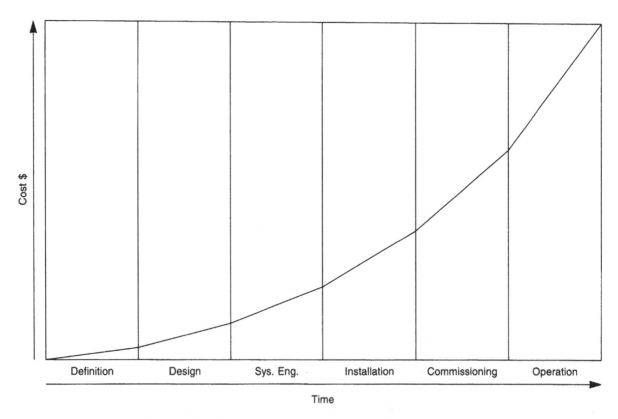

Figure 3-2. Cost increase escalation for changes at different project phases

jectives for the environmental control systems operations. These objectives are to:

1. Provide safe and comfortable environmental conditions throughout the building for year-round occupancy.

2. Provide an energy-efficient design which further stimulates energy awareness and energy conservation by the building occupants.

3. Select and install systems with operating parameters that assure continuous operation at designed parameters, minimum failure rate and maintenance requirements.

Program Development Objectives

Defining short-term objectives is aimed at meeting the following specific requirements and goals for the program development phase. However, another purpose of these objectives is to encourage participants to work as a team in meeting them.

1. Assemble a team of experts, and foster innovative ideas and openness, with the focus on established goals and objectives.

2. Solicit, evaluate, and incorporate input to the program definition from departments and organizations that have real-life experience in the operation and maintenance of similar systems and/or facilities.

3. Incorporate existing site design standards, specifications, and other site-specific requirements into the program definition.

4. Develop systems requiring minimum maintenance, and identify annual operating and maintenance requirements in terms of hours or dollars.

5. Develop solutions which lead to maximum utilization of natural resources and minimum energy use.

6. Recommend systems based on Life Cycle Cost (LCC) analysis rather than on first cost basis, and develop budget accordingly.

7. Define operating parameters for the building and individual spaces.

8. Provide documentation reviews and approvals by in-house and outside organizations.

Program Development Team

It is essential for the success of the project that the project manager assemble a program development team. This should be a team that can work well together, and try to achieve the set goals for the development phase of the project. The team should include the following members:

- Project Manager
- Commissioning Agent
- Building Users/Occupants' Representatives
- Facilities Engineers/Staff Architects
- Architects
- Consulting Engineers
- Systems Engineers
- Other Specialists

Program Development Review Team

The project manager and the commissioning agent should also establish a review process and assemble a review team. By announcing the review team, project management has established "customers" for the program development team. The review team should include the following members:

- Commissioning Agent
- Representatives of the Owner/Users/Tenants
- Facilities Engineers/Staff Architects
- Representatives of Operations and Maintenance
- Representatives of Budget and Finance Offices
- Value Engineering Team
- Other Agencies of the Owner
- Other Outside Agencies

In establishing teams, management should list team members by name to increase accountability and to foster good interpersonal relationships among the team members. While most of the members on the program development team are assigned to the project full time, design review is not a full time job. Review team members participate in scheduled meetings and documentation reviews as requested by the commissioning agent and the project management. Because their expertise is valuable, their time spent on reviews should be optimized.

Performance Requirements

A focus on performance should be the key requirement throughout the program development phase. In this phase, the building and its systems are being defined to meet design parameters and required environmental conditions. The design team should therefore consider building performance as one of the "program definitions" that the building and systems designs have to meet.

The disciplines that influence the building's future environmental conditions the most are architectural, HVAC- and BAS-related. Besides meeting aesthetical and functional requirements for the designed building, architectural design should also provide essential features that support the performance of the environmental control systems.

Architectural Definitions

The following architectural features have a profound effect on the environmental controls system's performance.

Building Envelope

- Design energy-efficient building for minimum energy consumption.
- Support economic control of basic building parameters such as temperature, relative humidity, pressure, light, and so on with minimum use of energy.

Windows

- Assure energy-efficient operations at design parameters.
- Design windows to assure minimum heat gain/loss, controlled daylight penetration, and minimum moisture condensation.

- Coordinate function with design parameters at extreme ambient conditions (temperature, relative humidity, air penetration, etc.), and with proposed HVAC systems (constant volume, Variable Air Volume, and perimeter heating).

Air Exhausts and Intakes

- Design their locations; size; use of grills, louvers, and dampers to preclude short cycling of exhausted air into the intake air.
- Preclude nuisance and environmental hazards (pollution) to the surrounding environment.
- Coordinate architectural features with the functional parameters specified by the HVAC engineer to provide specified air volumes at designed velocities, and prevent air penetration at closed positions.

Doors, Loading Docks and Other Large Openings

- Design to assure air locks which prevent loss of building pressure and penetration of outside air during adverse outdoor air conditions such as winter cold, summer heat, and high humidity.
- Design adjacent areas with additional heating or cooling to compensate for loss through large openings.

Space Allocation/Zoning

- Design spaces and interior zones adequate to the performed tasks.
- Zone and divide building to allow for temperature, humidity, and pressure control at their design parameters by conventional HVAC and BAS systems.
- Zone to allow for proper air balancing, to prevent migration of air from one zone to another.
- Match building zoning and HVAC zoning.
- For partial renovation of buildings (i.e., one floor), coordinate the zoning with adjacent spaces.

Mechanical Rooms

- Provide adequate space and conditions, to allow installation and servicing of mechanical and controls equipment.
- Separate mechanical rooms from occupied spaces to prevent disturbance due to traffic, noise, breakage, leaks, and repair of equipment.

Vertical and Horizontal Chases

- Design spaces for adequate vertical and horizontal distribution of air ducts, cables, pipes and other equipment such as VAV boxes, controllers, and so on.
- Provide for adequate service access and lighting, especially if there is mechanical or controls equipment located in such spaces.

Design of HVAC Equipment in Occupied Spaces

- Provide adequate space for installation of HVAC and controls equipment on occupied floors, such that their operation will be safe and will not disturb productivity and occupancy comfort.
- Design locations that allow for equipment maintenance without interruption of day-to-day activities in rooms, corridors, and other occupied areas.

- Choose equipment locations to minimize damage to the surrounding environment in case of equipment breakage, leaks, and other problems.

Service Access and Lighting

- Design adequate access for installation and maintenance of installed equipment.
- Lighting level at locations of DDC field controllers should be 30-50 foot candles (fc).

Local Conditions

- Meet the local conditions (cold winters in the North, hot, humid summers in the South—conditions that occur every year regardless of the average outdoor design conditions—see ASHRAE tables for towns in different geographical areas).

Mechanical Definitions

The following mechanical definitions should match those of the architectural features in the program definition:

Envelope Losses

- The HVAC design should compensate for all envelope losses by providing additional cooling, heating, or air locks for perimeter zones.

Zoning and Space Allocation

- Match zoning of the HVAC systems to zoning and space allocation.
- Provide for air and heat balancing with respective adjacent zones.

Adequate Ventilation

- Provide ventilation for a safe and healthy building environment, for all modes of operation. (Note: Complaints related to Sick Building Syndrome are due to inadequate ventilation and odors from construction and interior materials, sealants, glues and finishes. These are significant after building completion, when the tenants first move into the building.)

Supply and Exhausted Air Volumes

- Match supply and exhaust air volumes to provide desired ventilation and pressurization of individual rooms and zones.
- Prevent drafty environment and air contamination in the building.

Connections to Existing HVAC Systems and Utilities

- Coordinate the operation of new HVAC systems with existing HVAC systems serving adjacent areas in the same building.
- Coordinate connections of proposed new systems and equipment with existing utilities, such as electricity, chill water, steam and condensate return, and other on-site utilities.

Energy Efficient Systems

- Provide design of systems which meet site energy efficiency requirements.

HVAC Systems

- Meet site design performance and material standards.
- Design systems compatible with site operations and maintenance practices.

Controls and Automation Definitions

The following controls and automation-related definitions should be part of the program definition:

BAS System Architecture

- Hardware architecture, including specification of the front end Operator Work Station, its location and operating methods (i.e., by designated operators, building manager, etc.).

- DDC controller definition and its location (i.e., one controller per building).

- Unitary and application-specific controller specification and its allocation (i.e., one controller per HVAC unit).

BAS Software Architecture

- Software and software functions, control, data processing, data storage, and communications, at each level of the architecture.

Desired Degree of Automation and Energy Management Functions

- Functions such as set point reset by the operators, programs for HVAC system analysis, alarm analysis, and messages.

- List of energy management features.

Communications

- Between buildings (on LAN, dedicated network), using cables (telephone wiring, co-ax, fiber optics cables, etc.).

- Inside the buildings, via dedicated twisted shielded pairs, co-ax cables, thinNet (10BastT), and so on.

Interfaces

- To other (new or existing) BAS (methods of communication, protocols, identification of data for transfer, etc.).

- To maintenance management system (methods of communications, protocol, work entry, report generation, etc.).

- To other systems on the facilities network.

Systems and Network Security

- Identification of network nodes, and access levels, by area, systems, and functions.

Single Vendor or Multi-vendor

- Preferred arrangement for the facility.

- Methods of vendor qualifications.

- Responsibilities for systems interoperability.

- Long-term cooperation, partnership agreements.

- Service agreements.

Methods of DDC Systems Selection

- Bidding

- Pre-qualification

- Single sourcing

- "One" vendor's DDC for a building

Design Parameters for Environmental Control Systems

Design parameters, such as temperature, relative humidity, pressure, and so on, should be established for the building, and for individual spaces during the program development phase. Defined parameters can be uniform throughout the year, or can vary with seasonal changes and building occupancy. It is common to have different temperature, pressure, and relative humidity set points for summer and winter operations as well as occupied, stand-by, and unoccupied modes of operations. Building operating parameters (analog set points) can be easily changed in the DDC system by the operators or associated programs.

Performance Measures

Parameters for buildings and building systems' design are defined in standards, regulations, and design guidelines for individual engineering disciplines. Some facilities publish standards and design criteria to assure continuous uniform design, energy conservation, and standardization on material, procedures and performance.

The members of the review team "measure the performance" of the individual engineering solutions that have been submitted by the development team by evaluating their adherence to standards and other relevant criteria. Since individual team members bring to the team their own expertise (e.g., electrical engineering, HVAC engineering, DDC systems engineering, systems operation and maintenance, and financial), they are in the best position to evaluate the submitted documentation. Submittals should be of highest quality, and should contain drawings, specifications, calculations, expert opinions, approvals, and permits. The review team should be the final authority in the approval process.

The following is an example of a checklist that can be customized for specific jobs. Review team members can use the checklist to keep track of individual documents submitted for review, to request re-submittal in case a document needs additional work, and to rate the quality of the submitted documentation. The commissioning agent should use the checklists along with comments from individual team members for final evaluation of the program development documentation.

Identification of Customers in the Program Development Phase

In the program development phase, customer relationships are guided by contractual agreements. While this approach may be satisfactory for projects with minimum automation, projects with complex automation systems demand more interaction among individual participants, and between the design and review teams.

Many projects are streamlined if the owner's project manager hires an architectural firm, which, in turn, hires an engineering firm to provide the missing expertise. The approach works for the majority of projects. For projects with complex environmental controls systems, and for projects where the designed automation systems have to interface with other facilities' automation systems, the structure should be enhanced by informal customer relations.

To develop a realistic program for future systems operation at this early stage of systems conceptualization, the team needs input and skills from many different sources. Ideas and concepts should be discussed from different points of

Checklist for Program Definition Phase

Submitted Documents	Date of Submittal	Approved (A)	Not Approved (0)	Date of Resubmittal	Rating (Hi-10) (LO-1)
Architectural:					
Envelope					
Windows					
Exh/intake openings					
Doors/loading docks					
Space/zoning					
Mechanical rooms					
Vertical/horizontal chases					
HVAC equipment in occupied spaces					
Service access					
Lighting					
Design for local conditions					
HVAC:					
Design parameters					
HVAC alternative 1					
alternative 2					
alternative 3					
Zoning:					
Ventilation					
Supply vs. exh air					
Perimeter heating/cooling					
Interface with existing HVAC					
Electrical interface					
Chill water interface					
Steam/condensation interface					
Other interfaces					
Energy efficiency design					
System reliability					
Compliance with design criteria					
Compliance with site specs					
LCC cost alternative 1, 2, 3					
Best alternative					
DDC:					
DDC system for each alternative					
System architecture					
Software architecture					
System point capacity					
System communications					
BAS system interfaces					
System hardware					
System security/access					
Field gear					
Application software					
Operator interface					
Local support					
DDC unique features					
Interface to facilities network					

view with participants of different skills and knowledge. Such team effort goes beyond the boundaries of contractual obligations. The commissioning agent should break down the development phase into manageable (smaller) segments for the purpose of the review process. Each segment represents a document, and a design section made ready for review.

Most engineering reviews are best conducted in meetings addressing a particular design issue, and discussing it from different points of expertise. For example, a proposed HVAC system can be discussed by the architect (from the architectural point of view), the facilities engineer (from the facilities standardization, operational, and maintenance point of view), the systems engineer (from the systems controllability point of view), and so on. Each of these experts can enhance the HVAC systems proposal with his or her own expertise.

Professional and business relationships (in Total Quality Management) are perceived as customer relationships. Customer relationships are generally characterized by quality of goods or services, responsiveness to customer needs, courteous conduct of business—attributes that should also prevail among the team members. Regardless of whether the team members have contractual obligations with each other or informal relationships, their business contacts should be conducted as customer relationships. In the last checklist, reviewers become the "customers" to the HVAC engineer for that particular review segment regardless of formal contractual obligations.

Note that this review team, besides sharing opinions, should also have the final decision-making authority. As customers, and as experts in their own fields, they should be the final engineering authority to approve, or disapprove, certain concepts and solutions. They should also be responsible for the integrity of the proposed project when it comes to cuts due to budgetary constraints. No cuts (of any systems or their parts) should be made without consent from the Engineering Review Team. Project and facility managers should rely on the review team's engineering expertise for assurances that even the limited project (due to budget cuts) will provide all design criteria.

Figure 3-3 shows the segmentation of the program development phase, and the associated customer relations. The chart brings a look of formality, and gives an impression of added administration to the project. However, in the real situation, it does not have to be so formal. In fact, teamwork is best practiced in an informal environment. Many of the reviews can be done in sitdown meetings, or in the form of engineering briefings. However, the final documents should be formal and in compliance with the owner's requirements for submittal, design criteria, and site specifications.

To try to help the commissioning agent focus on pertinent issues associated with the quality assurance effort of the development phase, Figure 3-4 exemplifies key issues. The tasks are listed for the following phases: Definition, Planning, Implementation and Evaluation. Activities associated with each of these stages are project specific. The chart is a planning tool, and a reminder for the commissioning agent.

Future Automation Trends

To try to reduce operating costs, more and more facilities are turning toward the implementation of facilities automation systems. Facilities automation sys-

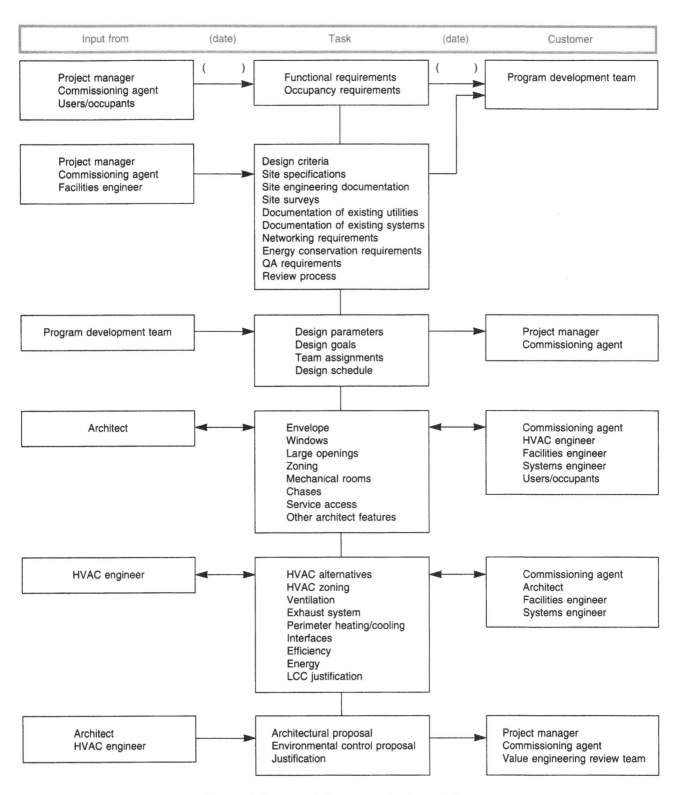

Input from	(date)	Task	(date)	Customer

Project manager
Commissioning agent
Users/occupants

( )

Functional requirements
Occupancy requirements

( )

Program development team

Project manager
Commissioning agent
Facilities engineer

Design criteria
Site specifications
Site engineering documentation
Site surveys
Documentation of existing utilities
Documentation of existing systems
Networking requirements
Energy conservation requirements
QA requirements
Review process

Program development team

Design parameters
Design goals
Team assignments
Design schedule

Project manager
Commissioning agent

Architect

Envelope
Windows
Large openings
Zoning
Mechanical rooms
Chases
Service access
Other architect features

Commissioning agent
HVAC engineer
Facilities engineer
Systems engineer
Users/occupants

HVAC engineer

HVAC alternatives
HVAC zoning
Ventilation
Exhaust system
Perimeter heating/cooling
Interfaces
Efficiency
Energy
LCC justification

Commissioning agent
Architect
Facilities engineer
Systems engineer

Architect
HVAC engineer

Architectural proposal
Environmental control proposal
Justification

Project manager
Commissioning agent
Value engineering review team

Figure 3-3. Customers in the program development phase

83

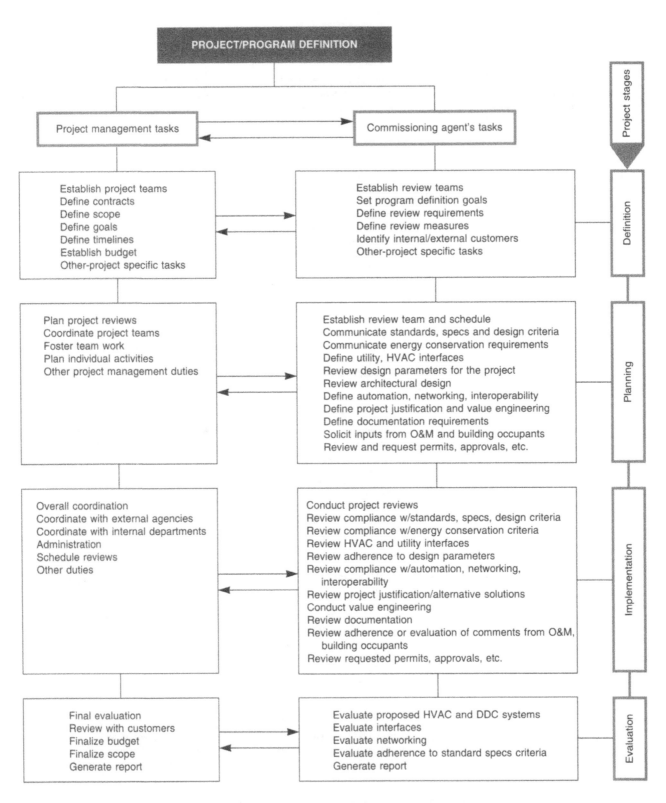

PROJECT/PROGRAM DEFINITION

Project stages

Project management tasks

Commissioning agent's tasks

Definition

Establish project teams
Define contracts
Define scope
Define goals
Define timelines
Establish budget
Other-project specific tasks

Establish review teams
Set program definition goals
Define review requirements
Define review measures
Identify internal/external customers
Other-project specific tasks

Planning

Plan project reviews
Coordinate project teams
Foster team work
Plan individual activities
Other project management duties

Establish review team and schedule
Communicate standards, specs and design criteria
Communicate energy conservation requirements
Define utility, HVAC interfaces
Review design parameters for the project
Review architectural design
Define automation, networking, interoperability
Define project justification and value engineering
Define documentation requirements
Solicit inputs from O&M and building occupants
Review and request permits, approvals, etc.

Implementation

Overall coordination
Coordinate with external agencies
Coordinate with internal departments
Administration
Schedule reviews
Other duties

Conduct project reviews
Review compliance w/standards, specs, design criteria
Review compliance w/energy conservation criteria
Review HVAC and utility interfaces
Review adherence to design parameters
Review compliance w/automation, networking,
    interoperability
Review project justification/alternative solutions
Conduct value engineering
Review documentation
Review adherence or evaluation of comments from O&M,
    building occupants
Review requested permits, approvals, etc.

Evaluation

Final evaluation
Review with customers
Finalize budget
Finalize scope
Generate report

Evaluate proposed HVAC and DDC systems
Evaluate interfaces
Evaluate networking
Evaluate adherence to standard specs criteria
Generate report

Figure 3-4. Overview of commissioning agent's tasks at project/program definition

tems—from BAS through maintenance management systems, administrative systems, and engineering systems—are undoubtedly the most cost-effective way of containing the rising costs of services.

While existing facilities struggle with the interoperability of their existing automation systems, new facilities are taking a global look at integration, beginning at the program development stage of their project. Real time automation systems (such as BAS, power plant controls and automation systems, production automation systems, utilities distribution controls, and metering systems) are being designed to operate on facilities' Local Area Networks (LANs) in coexistence with data management systems (such as administrative systems, inventory controls systems, maintenance management systems, and engineering systems). This integration allows the operating engineers to see the individual areas of operations, but they also provide a global look at other areas, for example, the energy utilization from production, distribution, and individual usage. A global view allows optimization of energy use and reduction of operating costs. Integration allows facilities managers to get management information (real time and statistical) from the systems on the network as they need them. With information distributed over the LAN, all parties connected to the network can access and share the information.

What does all this mean for the program development phase?

1. The owners and facilities managers need to develop master plans for the intended facilities automation.

2. Facilities automation for new facilities should be part of the program development phase. For existing facilities (with a good master plan and partial automation), the master plan should be reviewed by the program development team, then improved and adhered to in future projects. The program development phase should have a global view on long term development of facilities operations. Development of facilities automation should be part of this project phase.

3. Project development teams, architects, and engineers need to understand the future trends. Project managers should include operations experts, automation systems engineers, integrators, telecommunications engineers, data processing experts, and other non-traditional disciplines in the traditional project development teams. The complexity of automation systems, their integration, and the benefits to the owner, warrant expansion of the traditional teams into the new expertise.

Chapter 4

Design Phase

From Traditional Design/Bid to More Enhanced Systems Engineering

In the design phase, the focus should be on designing the specified HVAC systems, and specifying the controls and automation (DDC) systems for the project. This is assuming that selection and justification of the most appropriate HVAC system and definition of the DDC system requirements were concluded in the program development phase of the project. If the design phase was not preceded by a program development phase, the engineering team should develop HVAC design alternatives, and define the DDC system requirements at this phase of the project. Design alternatives should be based on a thorough site survey, interviews with the owner's O&M personnel, and with input from systems engineers. This is especially important for retrofit jobs of existing facilities, partial building renovations, and for sites with existing Building Automation Systems (BAS).

In "traditional" design/bid projects, the HVAC engineer coordinates the design with other engineers, such as electrical, piping, plumbing, fire protection, and other engineers. Most design teams are without systems or DDC applications engineering expertise. The situation has it roots in the simpler HVAC controls of the past. Usually, the simple single loop pneumatic controls hardware was specified by mechanical engineers, and the controls hardware was considered part of the mechanical (HVAC) equipment. For example, with the design of a dual duct air handling unit, the HVAC engineer also specified four single loop controllers (one mix air, one discharge air, one cold and one hot deck controller). Mechanical contractors bidding on the job treated the controls as any other fully designed and specified component of the air handling unit. They then proceeded with requests for bids from their subcontractors, including controls vendors, to receive competitive prices.

Because the controls subcontractor responded to the same design documentation as, for example, the sheet metal sub-contractor, and had the same bid requirements (i.e., sizing of equipment, pricing), all subcontractors had equal conditions for bidding in the allocated time frame. (See Figure 4-1 for an example of a design/bid timeline for simple controls.)

These past practices are carried over to the present, despite the complexity of modern Building Automation Systems. Without applications engineering expertise on the design team, and acknowledging that Building Automation Systems are far more complex than the single loop pneumatic controllers used to be, the systems engineering work has to be done by the controls subcontractors during the time frame allocated for bidding.

In this section, consider the activities of subcontractors during the time frame allocated for bid preparation. (See also Figure 4-2 for the major tasks of subcontractors during bid preparation.)

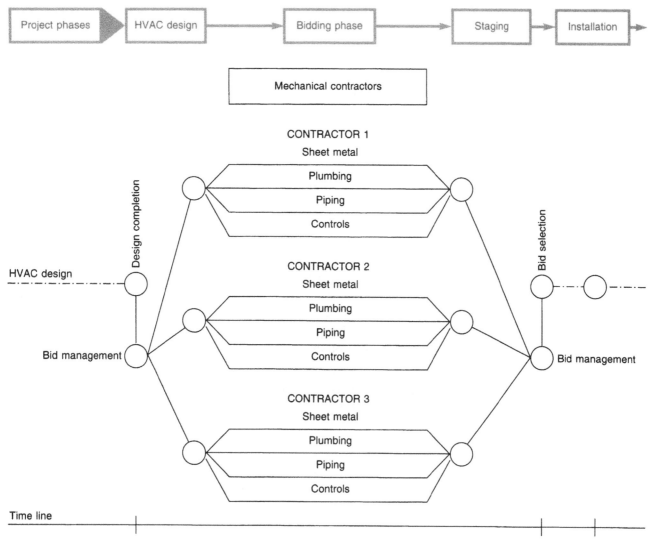

Figure 4-1. Design/bid timeline for simple controls

Mechanical and DDC Subcontractors' Tasks

The mechanical subcontractors, basing their work on "take offs" from the engineering HVAC documentation (equipment specifications, equipment locations, piping and duct work layouts, etc.), compile quotes for major equipment, prepare price information for material, labor, mark-ups, discounts, and so on.

The DDC subcontractors, upon receiving the same (HVAC) engineering documentation, have to design and then price the DDC system. To accomplish this, the DDC system engineers have to:

1. Review the HVAC drawings and point list (if provided) to determine the field hardware (number and types of field points)

2. Design one line control schematics based on the HVAC design and point list

3. Specify individual DDC controllers (building- and application-specific) according to individual control schematics

4. Determine DDC hardware architecture

5. Specify outside purchased hardware products (valves, sensors, actuators, transformers, etc), and major equipment (variable frequency drives, metering systems, etc.)

6. Prepare documentation for cost estimation of electrical wiring, controls field hardware installations, etc.

7. Review (or survey) requirements for communications and networking

8. Determine the most suitable means of communications within the building, between building controllers and operator interfaces, to other (third party) controllers, and other devices on the network

9. Review the HVAC engineers' sequence of operation, determine software requirements for application software, and review specified energy management features. Define interface software to third party controllers (drivers, point mapping, and data presentation), and interface software to other network devices (communications protocols, point mapping, data handling, data storage, and data presentation) if such devices and requirements are specified by the engineer.

10. Determine operator interfaces, and advanced features on the Operator Work Stations (OWS), such as colorgraphics, data trending, report generation, interfaces to third party software, and other features, software, and programming tasks on the OWS.

11. Prepare documentation and compile quotes for third party purchases (equipment, communications, installation), prepare price information for material, labor, mark-ups, and discounts, and submit the bid documents to the mechanical contractor.

Figure 4-2 shows that:

    a. Building Automation Systems engineering is being done after completion of the design work on the project.

    b. Systems engineering cannot add value to the project by modifying (simplifying) engineering solutions at the bidding stage.

    c. The time allocated for detailed systems engineering (usually 4 weeks) is insufficient for engineering a true automation system.

Insufficient engineering time and the desire of controls vendors to get the job (being competitive by bidding the lowest cost system) results, in many instances, in a minimal (DDC) system while meeting the HVAC engineer's specifications. This situation demonstrates itself in follow-up changes during installation and frequent upgrades of installed DDC systems. Unfortunately, many such DDC applications provide only little more than the single loop pneumatic controllers, for a much higher cost to the owner.

The Most Common Consequences of the Present Practices

1. If the lowest bid price is the criteria for selection, the lowest-cost DDC system is selected for the job. The systems are often "bare bone" systems, in compliance with the engineering specification and sequence of operation. Since most sequences of operation are written by HVAC design engineers (without consultation with systems and/or facilities engineers), they relate to control of the associated HVAC systems rather than to building automation.

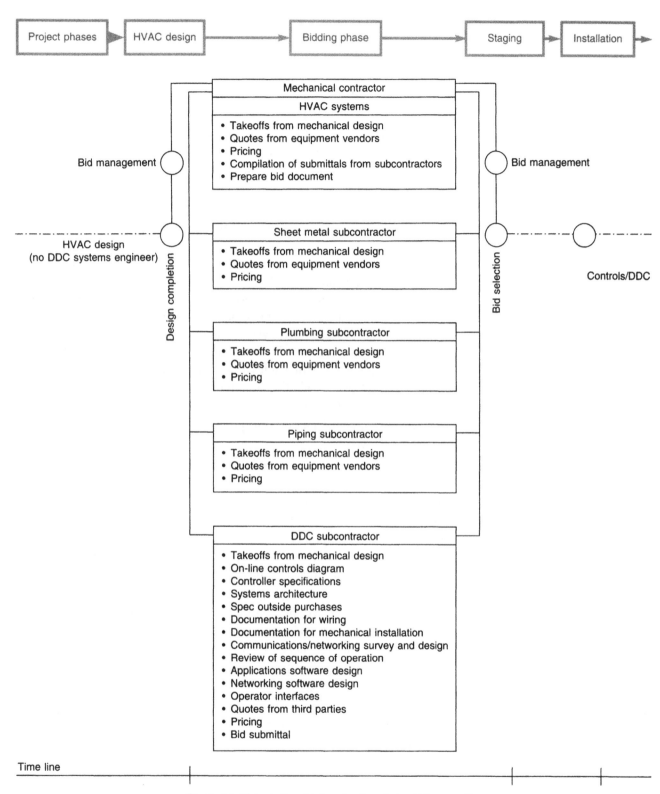

Figure 4-2. Major tasks of subcontractors during bid preparation

2. Because of the limited time to prepare proper DDC bids, there usually isn't time to coordinate the mechanical design with systems engineering. This results in the design of generic systems with minimal controls and automation features. Since the environmental controls systems performance is a function of both HVAC and DDC systems, the approach may result in substandard systems performance.

3. Advanced controls features available with most DDC systems are excluded from the bid for one simple reason—they increase the bid price due to additional engineering and applications software development costs.

4. There is the possibility of over-design of HVAC systems and over-specification of the controls requirements by the mechanical engineer. Without the benefit of consultations with the systems engineer and facilities engineer, HVAC engineers tend to over-design their systems to provide for worst case scenarios.

5. Because of insufficient time for proper DDC systems and applications software engineering during the limited time for bidding, some deficiencies surface during the installation phase, resulting in engineering change orders and additional costs. Other deficiencies, especially deficiencies related to applications software and systems architecture, surface during regular systems operation, resulting in additional expense to the owner in the form of systems upgrades. Deficiencies related to communications interfaces (or lack of them) result in additional systems engineering work to provide the owner with networked, interoperable systems.

6. Insufficient time for DDC systems proposal review by the HVAC engineer and facilities engineer may result in lack of compliance with site-specific standards and operations and maintenance requirements.

7. Limited or no time for pre-ordering of controls hardware that require long deliveries (6–8 weeks) may result in having to pay a premium for such orders, in order to meet the job schedule.

8. Insufficient time for proper selection of third party systems for specific applications, or to consider options related to communications, interfaces, and other site conditions creates further problems.

9. Without requesting bids from the pre-selected DDC vendors only and competitively bidding every job, the owner may end up with several incompatible DDC systems (and networks) on the site, and with increased training, operating, and maintenance costs.

10. Project deadlines may lead to evaluation of bids on the first cost, rather than Life Cycle Cost basis.

Alternatives to Present Design and Bidding Practices

1. Pre-select qualified DDC vendors. This provides the facility with DDC systems that meet the owner's requirements and the site's conditions, prior to individual projects. The pre-selection process disqualifies DDC vendors whose systems do not comply with site requirements, or who lack local systems support. The owner can standardize the pre-selected systems, train the O&M departments and assure systems interoperability at a reasonable cost.

2. Bid DDC systems when 75% of the HVAC design is complete. At that stage, most HVAC components are defined and specified. The HVAC design can then be completed with a particular (selected) DDC system in mind. This allows the engineers to design true environmental control systems for maximum performance and with DDC and HVAC components fully integrated. Deviations (from the original proposal at 75% completion) related to changes of the HVAC and DDC system can be managed on a per unit cost basis (i.e., unit price for an installed room temperature sensor, $/hr of application or software engineering, and so on). However, experience shows that such cooperation will result in overall cost reduction for HVAC and DDC systems. This can be attributed to the optimum design of both systems rather than design for worse case scenarios.

3. Provide ample time (and money) for systems engineering, and systems integration in the design phase of the project. In an effort to reduce operating costs, facility managers are looking more and more toward automation and integration of systems. Project managers should budget for automation and integration design in their projects. It is a case of "pay now or pay later" with a three to five times escalation cost in the later payments.

4. Design systems with a higher degree of automation. DDC systems have capabilities not only to control the connected HVAC system, but also to provide a higher level of automation, such as:

- Executing more complex control algorithms

- Assisting the operators with problem analysis

- Executing programs associated with self tuning, calculation of start times, set points, and analysis of HVAC related problems

- Providing advanced Person Machine Interfaces (PMI), in the form of graphical interfaces, trend graphs, report generation, and export of data into common software media, such as spreadsheets and drafting software

- Providing interfaces to third party controllers designed for specific applications

- Providing data transfer over a facility network to other systems, such as a maintenance management system

- Providing interfaces with real time systems, such as power plant and utilities distribution systems, and providing the owner with advanced automation and management tools.

5. Project managers should foster good working relationships of in-house engineers with outside (HVAC) engineers, and vendors' system engineers during the design phase of the project. This is important for optimum systems design, adherence to site standards, and interfaces with existing systems.

6. Justify projects on a Life Cycle Cost basis rather than on a first cost basis. The first cost can be deceiving, especially with automation systems, where over 50% of the expenditure is for engineering and software development. It is very difficult to detect engineering and software deficiencies during the traditional bid evaluation process. Focus on performance, familiarity with the automation system, and a structured review process should be implemented to safeguard the process. Engineering, hardware, and software items aimed to reduce operating and maintenance costs should be included in the design.

Even though these items may increase the initial cost for the project, in the long run they are probably the most cost effective measures for the owner.

Applying the Commissioning/TQM Approach to the Design Phase

The commissioning agent should develop the three basic TQM measures for the design phase of the project:

1. Define design objectives and assemble teams
2. Establish performance requirements and measures
3. Identify customers

Design Objectives

The design of environmental control systems (HVAC and DDC) must comply with the chief objectives listed in the Performance Specification:

To provide a complete and fully operational environmental control system that interfaces with existing systems on site, and is installed in accordance with current national, state, and site-specific standards and specifications.

To meet this objective, the commissioning agent and other participants should concentrate on the following goals:

1. To solicit, evaluate, and incorporate input relevant to environmental control systems from the owner's departments, as well as from outside organizations that have real-life experience in the operation and maintenance of similar types of facilities.

2. To comply with the design review process and incorporate review comments into the HVAC and DDC design.

3. To incorporate site Design Standards, Performance Specifications, and other site-specific documentation into the design documentation.

4. Honor existing site-specific requirements for standardization of design, documentation, equipment, material, HVAC and DDC systems, and communication methods already in place at the owner's site.

5. Design systems compatible with operating and maintenance practices of the owner, and identify initial training and annual operating and maintenance requirements.

6. Design systems which lead to maximum utilization of outside air, and recovery of heat from exhausted air.

7. Comply with site energy standards, and use outside financial resources, such as energy credits, tax breaks and rebate programs.

8. Provide properly engineered interfaces to site utilities, and to existing HVAC, controls, and other systems.

9. Utilize fully the DDC system features.

10. Develop a Sequence of Operations that will assure a building environment at designed parameters year round, with safe and energy efficient operation at minimum maintenance cost.

11. Provide complete documentation of highest quality and standards for each of the review stages, as requested in the Design Standards and Site Performance Specifications.

Teamwork in the Design Phase

Fostering good relationships and teamwork within individual design teams and among the teams is the key to a successful project. A good team can solve many complex problems and overcome many obstacles during the design stages of the project. In new buildings, design problems most often arise from conflicting requirements, budgetary constraints, incomplete coordination of engineering disciplines, and other design issues. In retrofit installations, these problems are further aggravated by having to match the new design to existing building and/or environmental control systems. In both types of projects, coordination and teamwork are great challenges for the project manager and commissioning agent.

Design teams are assembled from architectural and engineering organizations that were selected for the project. Highly reputable firms are often national or international with offices remote from the job site. The project management wants to be sure that the focus of the individuals is on productivity, and delivering the best design on time, as specified in the contract.

This concept is very productive and efficient for new construction projects that have good program definition and a well-estimated budget. For renovations and retrofit installation, the design team has to go beyond program definitions to consider site-specific conditions and budgetary constraints. Not all facilities have perfect documentation of their existing installations and distribution systems. Establishing a good working relationship between the design team and the owner's departments (facilities engineers and O&M personnel) can overcome many unforeseen problems.

Besides good teamwork, a focus on the quality of future systems operation is also the key to a successful project. This should be considered by the project manager and commissioning agent from the beginning of the project. A good project manager should draw on both the professional qualities and resources of the design team, as well as in-house expertise. The project manager should balance available in-house expertise with design services provided by consulting engineers to achieve optimum utilization of intellectual and financial resources.

Case Studies

The importance of good teamwork is illustrated in the following two case studies.

Case Study 1

The design team for a new building was managed by an architect who was one of the directors of the facility. The design was completed on time, with beautiful architectural features, and with environmental controls systems. The project was delivered to the selected construction management firm for estimating. The estimated construction cost was higher than the project budget (to everybody's surprise). It was time to make sacrifices; time to modify the design, cut costs, and fit the project within the budget. Reducing costs by doing away with some of the architectural features (eliminating expensive wooden paneling, architectural lighting, and so on) was out of the question. The spotlight was turned on the reduction of the scope of environmental controls systems.

After reviews and meetings, the HVAC engineer was persuaded to revise the design and "engineer out" the hot water perimeter radiation heating from the offices on the southern exposure of the building. It satisfied the design team because the offices had a southern exposure. (By the way, who would argue with a director who awards all future contracts for such a large company that has ongoing construction activities?) The project received an energy rebate award for implementation of Variable Frequency Drives (VFD), DDC control of VAV units, and implementation of energy management features.

The site was in the northern United States. The construction was completed on time, DDC controls installed, HVAC systems balanced, and the building was turned over to the users. The happiness with the new systems lasted until the winter day that temperatures dropped to the 20s, and night temperatures to the teens. They stayed there stubbornly for weeks. Calls to facilities operations kept coming in. The fan discharge air set point was increased far over the heating coil's capacity, and so was the VAV system's air velocity. The building start time was pushed back, then the night setback was eliminated altogether (forget energy conservation!). Now, the employees were not only cold, but they also complained about air velocity (drafts) and increased noise level.

Because the building had been completed in the early spring, the designers, decision makers, and contractors were gone. It took several weeks before the operating and maintenance department realized that the south side of the building did not have perimeter radiation. As the winter faded away with increasingly warmer temperatures, there were fewer and fewer complaints, and by the spring, operation was back to normal. As in many such cases, the building's occupants and the operations and maintenance departments had to learn to cope with the building deficiency. It was that, or face up to an unpopular decision to renovate the heating system in a perfectly new building.

In this case study, the decision to "engineer out" the perimeter radiation heating system was influenced by the director in favor of preserving proposed architectural features. In the absence of a design review team with final authority for design approval, the decisions are often influenced by other than engineering considerations.

Case Study 2

The dual duct air handling units in a twenty-year-old laboratory building were up for renovation and for new DDC controls. The survey showed that the ducts were clean and in good condition, and the AHU's coils were recently replaced. The building needed new DDC controls, replacement of outside air dampers, humidification and de-humidification, coordination of supply and exhaust air, fume hood controls, and other mechanical upgrades. The original idea was to convert the dual duct air handling system to a Variable Air Volume system. The job estimate was far over the budget.

The project manager assembled a team of engineers consisting of an HVAC engineer, systems engineer, air balancer, and a facilities engineer. The team took another look at the building and spent a lot of time talking to the building's occupants and to laboratory researchers. The process did not look too structured, and not many memos and directives were written and issued (nobody felt the need to cover their behinds). Instead, a lot of engineering discussions and confrontations took place. An engineering survey of the existing conditions of mechanical and controls equipment was completed along with a survey of individual laboratories (and their operation).

It was decided to renovate all major equipment associated with environmental controls systems, and to retrofit existing lighting with new energy-saving ballasts, lamps, or fixtures. The goal of the project was to provide the building with fully operational and economical environmental control systems and lighting that required minimum maintenance. With the combination of HVAC, controls, and lighting retrofits, the energy simulation proved a good return on investment. The project qualified as an energy project, and qualified for the utilities rebate program.

The design solution was to convert the existing AHUs to VAV units and to install VFD drives. With minimum interruption to the building's operation, new VAV boxes (with flow measuring stations) were installed into both hot and cold ducts, before the existing mixing boxes. The interiors of existing mixing boxes were removed, and pneumatic room thermostats were replaced by DDC sensors with local set point adjusters and night setback override buttons. Each pair of VAV boxes got a DDC controller. The supply air into the laboratories was matched with the general and fume hood exhaust air, to create slightly negative pressure in each laboratory. To provide energy-efficient operation and savings due to night setback, the night temperature was set back by 5–10°F, and the air volume was set back to a 65 fpm air velocity (measured at the face of the worst case fume hood). The challenge for the controls technicians and air balancers was to balance each fume hood, find the lowest air flow, and balance the entire system accordingly.

The retrofit of air handling units was done with minimum interruption of air delivery. Modifications of mixing boxes and installation of VAV boxes in the laboratories (about 4 hours per laboratory) was always scheduled ahead and coordinated with the researchers. The payback on the project was less than 4 years; metered electrical energy savings, compared to the last 4 year average were 25–30%. The building temperature, humidity, and air flows were well balanced throughout the offices, laboratories, corridors and other spaces in the building. The environmental control system turned out to be one of the "quietest" anybody could remember in a twenty year old building. Because the project came under budget and there was money in the contingency, it was decided to use the money for other deferred maintenance items, such as fume hood repair, steam and condensate system repair, and other items identified by the maintenance department, environmental safety officer, and the laboratory users.

This case study is an example of a well-assembled team of engineers. In their decision-making, ideas rather than rank were important. The team focused on the performance of designed and renovated systems, and how to meet the individual needs of the laboratory researchers.

How to Select a Good Team

These two case studies demonstrate the importance of well-defined and managed design teams. They show that the project manager and the commissioning agent should select team members capable of stimulating ideas, and engineering discussions, and designing systems which meet the owner's performance requirements.

Design Review Checklist: 30% Design Completion

Submitted Documents	Date of Submittal	Approved (A)	Not Approved (0)	Rate of Resubmittal	Rating (Hi-10) (Lo-1)
Compliance with design criteria					
Compliance with site specs					
HVAC schematic					
HVAC system layout					
Equipment sizing					
Calculated temps, RH, air					
HVAC preliminary spec					
Building zoning					
Ventilation					
Heat recovery					
Free cooling					
Energy design					
Energy rebates/incentives					
Utilities interfaces					
Coordination with other HVAC					

Design Review Checklist: 75% Design Completion

Submitted Documents	Date of Submittal	Approved (A)	Not Approved (0)	Rate of Resubmittal	Rating (Hi-10) (Lo-1)
Compliance with design criteria					
Compliance with site specs					
Equipment sizing					
HVAC schematic					
HVAC system layout					
Calculated temperatures, RH, air					
HVAC job specification					
Service accesses					
Zoning					
Ventilation					
Heat recovery					
Free cooling					
Utilities interfaces					
HVAC interfaces					
Energy efficient design					
Energy rebates/incentives					
DDC point list					
Sequence of operations					
DDC Interfaces					
Networking					

Design Review Checklist for Completed Design

Submitted Documents	Date of Submittal	Approved (A)	Not Approved (0)	Rate of Resubmittal	Rating (Hi-10) (Lo-1)
HVAC one-line diagram					
HVAC system layout					
Equipment sizing					
Compliance with design criteria					
Compliance with site specs					
Calculated temps, RH, Air					
HVAC job specification					
Service accesses					
Zoning					
Ventilation					
Heat Recovery					
Free Cooling					
Air Distribution design					
Hydronic System design					
Room Pressurization					
Connection to utilities					
Interface to other HVAC					
Energy design					
Energy rebates/incentives					
DDC point list					
Sequence of operations					
DDC architecture					
I/O devices					
Controllers					
Hardwired interfaces					
AC supplies					
OWS					
Systems features					
Third party controllers					
Networking					

Selection of design teams and individual team members should be based on good management practices and common sense approaches. The following should aid the project manager and the commissioning agent in the selection process:

- In addition to the design team, establish informal problem solving teams, assembled of professionals whose business is related to the defined task.

- Each member of the team should be an active participant, a problem solver, who can contribute to the task at hand.

- Define objectives, and make sure that each member is committed to contributing to and reaching the objectives.

- Avoid people whose only job is to attend meetings.

- Avoid ranks—ideas and solutions rather than management ranks should count.
- Decline voting privileges and decision-making to people who have limited understanding of the (engineering) subject being decided on, even if they are managers or are on the project team.
- Invite people who can even remotely contribute to the subject from different angles; for example, for decisions on laboratory environmental controls, invite professionals who can contribute to the design solution (such as fume hood experts, fume hood controls experts, chemical safety experts, air balancing engineers, and most of all, laboratory researchers).

The following is an example of a design team with a focus on environmental controls systems:

- Project Manager
- Commissioning Agent
- Architects
- Structural Engineers
- Mechanical/HVAC/Electrical Engineers
- DDC Systems Engineer
- Other Consultants
- Specialists
- Job Estimators

Combining such diverse talents and interests in one group is appropriate for project coordination, but less appropriate for decision-making and approval of specific engineering design solutions. The consensus in such a large group may lead to a decision that is not technically sound. Meetings can also be very time consuming due to the number of attendees and variety of topics being discussed.

In the design phase of environmental controls systems, the focus is shifting from global to specific engineering tasks. Early selection of the controls system from the pool of pre-selected systems, and participation of the DDC systems engineer in the process, is appropriate for good HVAC design. Participation of DDC systems engineers on the design team should start no later than at 75% of HVAC design completion. Participation of systems engineers in the design or project review process will save substantial costs for HVAC as well as DDC systems.

The Review Team

The environmental control system design should be reviewed at, for example, 30, 75, and 100% of design completion by a team assembled by the commissioning agent. Recommended members of the team are:

- Commissioning Agent
- Facilities Engineering
- O&M Personnel
- Energy Manager
- Users/Occupants
- Other Specialists, as necessary

The review process should focus on quality of design, and also on meeting Performance Requirements, as established in the Program Definition, Design Standards, and site Performance Specifications.

Issues and changes related to environmental control systems crop up during the design stage of the project. Such issues may be related to modification of the original program definition, or architectural or budgetary changes. Engineering changes should be resolved by the commissioning agent and the engineering team capable of evaluating the overall impact on the project. For example, the impact of elimination of a perimeter heating system from the project should be evaluated by the team led by the commissioning agent, consisting of the HVAC engineer, facilities engineer, systems engineer, and the O&M manager responsible for future systems operation and maintenance. For specific applications, there may be a need for other specialists. For example, for issues related to laboratory environmental control systems, the team should include an industrial hygienist or a representative from the facilities chemical safety department.

Performance Measures

The commissioning agent should develop review and evaluation procedures, checklists, and related forms for different stages of design review. The performance measures used should relate to the reviews of design documentation, specifications, and engineering calculations. These are reviews that check on systems performance, compliance with site design standards, specifications, codes, and other standards. The Quality Assurance process at this phase of the project depends more on the expertise of the review team than on an analytical process.

The following examples of review checklists show the items most commonly checked for individual design-completion. As in the preceding phase, the checklists should be modified by the commissioning agent for the project and site conditions.

Identification of Customers

The commissioning agent should identify customers for every significant task in the design process. Because every project is unique, Figure 4-3 serves only as an example to identify major tasks and relationships among the design team members and individual tasks.

Applying the Commissioning/TQM Approach to Design/Build Projects

HVAC or DDC systems retrofit projects are often implemented as design/build projects. Such projects do not require re-engineering of the building or of its mechanical systems, but they are an upgrade of the mechanical and/or control systems. The design and engineering of such projects can be limited to HVAC and/or controls design. Some retrofit installations may require architectural design, especially if they include changes of space usage or layout. For design/build projects, the number of participating engineering organizations is greatly reduced and the project management is more straightforward.

Many energy projects are design/build projects. Major controls vendors as well as mechanical contractors specialize in turnkey projects, providing the

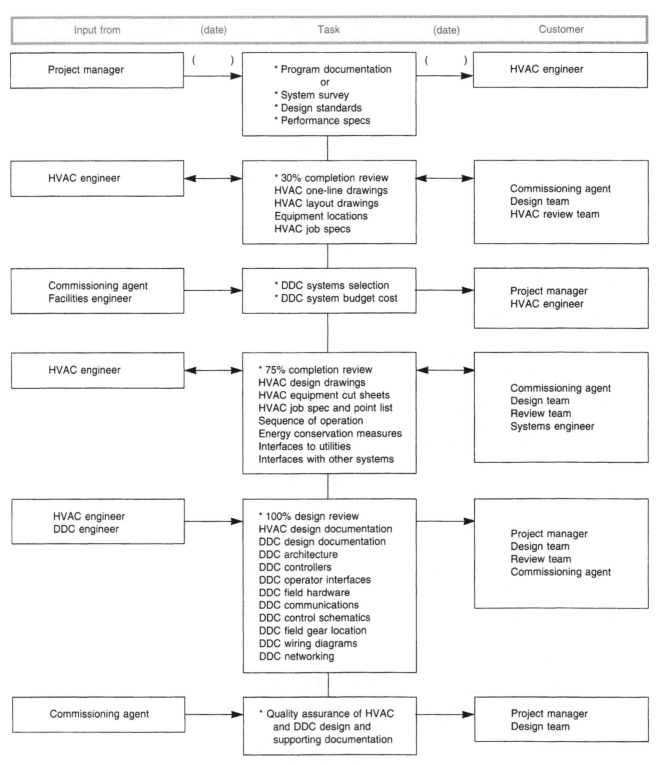

Input from	(date)	Task	(date)	Customer

Project manager → ( ) → * Program documentation
or
* System survey
* Design standards
* Performance specs → ( ) → HVAC engineer

HVAC engineer ←→ * 30% completion review
HVAC one-line drawings
HVAC layout drawings
Equipment locations
HVAC job specs ←→ Commissioning agent
Design team
HVAC review team

Commissioning agent
Facilities engineer → * DDC systems selection
* DDC system budget cost → Project manager
HVAC engineer

HVAC engineer ←→ * 75% completion review
HVAC design drawings
HVAC equipment cut sheets
HVAC job spec and point list
Sequence of operation
Energy conservation measures
Interfaces to utilities
Interfaces with other systems ←→ Commissioning agent
Design team
Review team
Systems engineer

HVAC engineer
DDC engineer → * 100% design review
HVAC design documentation
DDC design documentation
DDC architecture
DDC controllers
DDC operator interfaces
DDC field hardware
DDC communications
DDC control schematics
DDC field gear location
DDC wiring diagrams
DDC networking → Project manager
Design team
Review team
Commissioning agent

Commissioning agent → * Quality assurance of HVAC
and DDC design and
supporting documentation → Project manager
Design team

Figure 4-3. Identifying customers for tasks in design process

owner with surveys, designs, project justification, installation, and in some cases, with financing of projects based on energy savings.

In design/build projects, the role of design firm and general contractor is assumed by the controls (DDC) or mechanical contractor. The success of these projects greatly depends on participation of the in-house resources (facilities engineering, O&M departments) with the outside organizations (DDC and/or mechanical contractors).

Retrofit installations done jointly by these groups are usually successful because both organizations have long-term interest in the operation of the environmental controls systems. Many contractors selected for design/build projects have had long-term involvement as well with the facility in their roles as engineers, contractors, or service organizations. These long-term relationships result in the contractor's familiarity with the owner's facilities and O&M procedures, which in turn results in mutual trust, and a feeling of ownership for the project.

Another reason for success of design/build projects is increased focus of the participating organizations on long-term operations and maintenance of the retrofitted systems.

Because many DDC and/or mechanical contractors are local, they are familiar with the existing systems, operation, and maintenance practices. Site surveys, the most important phase of retrofit projects, are conducted with a previous knowledge of existing mechanical systems, equipment, utilities systems, and controls. Besides doing site surveys, the contractors tend to spend more time on site, interviewing customers and O&M personnel. This leads to development of a good working relationship between the contractor and the owner's organizations. The relationship is very important for work scheduling because most retrofit installations are done with only partial disruption of building activities.

Fostering good relationships with the building's occupants means meeting with them, and informing them of daily activities in the building and in their individual areas. By recognizing and understanding the daily routines of departments or individual offices, the contractor can schedule the construction activities so as to cause minimal disruption to building operations.

The success of engineering and implementation of retrofit projects by the participating groups also depends on their engineering and project management capabilities. It would be a mistake to automatically assume that a good service organization can provide good engineering and design services. Larger, more complex jobs require extensive engineering and project management expertise. Underestimating these requirements could result in installation delays, budget overruns, and inadequate engineering and job documentation. Best results can be achieved by complementing the resources of the outside DDC and/or mechanical contractors with in-house engineering and project management resources. A well-appointed team, managed by an in-house project manager or commissioning agent, can provide most design and project management related services for retrofit jobs.

Design/build projects are part of the daily construction and retrofit installation landscape. They range from small construction projects to high tech specialty projects. In the facilities industry, the owner has the choice of contracting an HVAC and/or controls retrofit design/build job with a controls contractor, a mechanical contractor, or an energy contractor. Facilities automation systems integration jobs can be contracted with the participating DDC vendors, a systems integrator, or with systems houses specializing in facilities automation and networking. A power plant retrofit job can be contracted with the service organization providing related services, major chiller or boiler manufacturers, firms specializing in power plant (or industrial automation), or a systems integrator.

Checklist for Design/Build Projects

Submitted Documents	Date of Submittal	Approved (A)	Not Approved (0)	Rate of Resubmittal	Rating (Hi-10) (Lo-1)
Building survey					
HVAC systems survey					
BAS survey					
Air balancing survey					
Survey of mechanical systems					
Survey of electrical systems					
Survey of utilities interconnections					
Survey of communications					
Survey of existing steam/condensate systems					
Survey of chill water distribution systems					
Survey of hot water systems					
Survey of building occupants					
Survey of operations					
Survey of maintenance					
Compliance with design standards					
Compliance with performance specifications					
Determine existing systems' reliability					
Proposed utilities interfaces					
Energy efficient design					
Energy rebates/incentives					
HVAC one-line diagrams					
HVAC system layout					
Equipment sizing					
Calculated temps, RH, air					
HVAC job specification					
Zoning					
Ventilation					
Heat recovery					
Free cooling				.	
Air distribution					
Hydronic system					
Room pressurization					
Interface to other HVAC systems					
DDC point list					
Sequence of operations					
DDC system architecture					
I/O devices					
Controllers					
Hardwired interfaces					
AC supplies					
Operator work stations (OWS)					
OWS features					
Application programs					
Interface to third party controllers					
Networking and communications					

Performance Measures for Design/Build Projects

In a design/build project, the tasks of the commissioning agent are similar to those in the design phase of a design/bid project (see page 104). However, engineering tasks should be always preceded by a thorough site survey.

The following is an example of a checklist to aid the commissioning agent setting up Quality Assurance measures for design/build projects.

Identification of Tasks

In design/build projects, assigning tasks and forming customer relationships are greatly simplified because long-term relationships are already established. Table 4-1 shows these inter-relationships.

Figure 4-4 is an overview of the commissioning agent's tasks in the design phase of a project. The primary focus is on adherence to design criteria and standards, quality of site surveys, interfaces to existing utilities, air systems, and DDC communications interfaces. DDC system and vendor selection, as described in the following chapter, should be conducted before, or during, the design process. This assures having pre-selected DDC vendors ready for bidding (or for other means of systems selection), before 75% completion of the HVAC design.

Tasks	Executed by
Project management	Owner's project manager and/or contractor's project manager.
Commissioning agent	Owner's plant engineer
Project definition	Owner's plant engineers
Engineering survey	Plant engineers O&M Contractor
Review of survey documentation	Plant engineers O & M Occupants
Project justification Approval	Plant engineers O&M management Facilities budget office
Engineering	Contractor Plant engineers
Engineering documentation approval	Plant engineers O&M
Installation	Contractor
Commissioning	Contractor O&M Plant engineers
Turn over & training	Contractor O & M Plant engineers
Warranty operation	Contractor

Table 4-1. Inter-relationships of Participants in Design/Build Projects

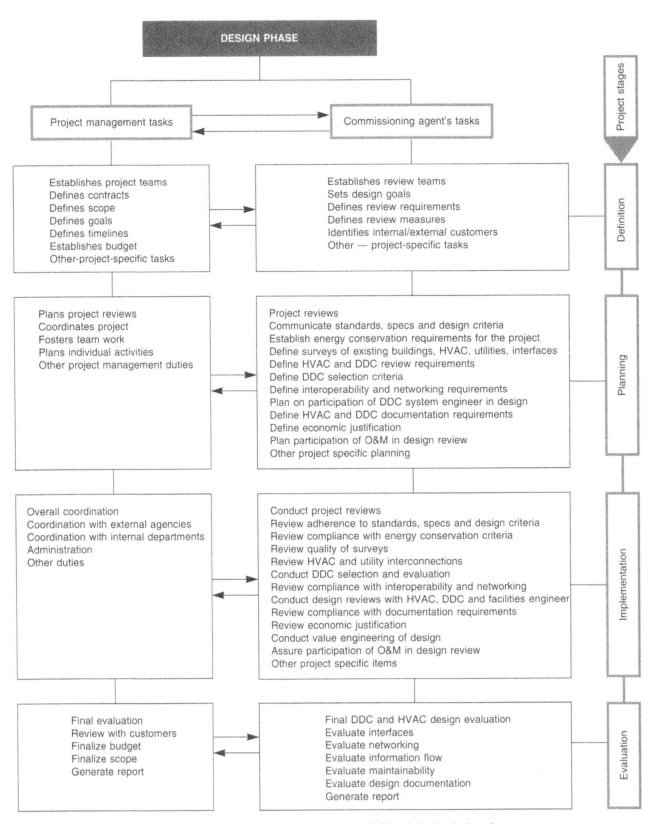

Figure 4-4. Overview of the commissioning agent's tasks in the design phase

Following the DDC system selection, the review team and the commissioning agent should focus on:

 a. Systems engineering of the HVAC and DDC systems into an integrated "environmental control system".

 b. Interoperability (networking, interfaces, data transfer over a common network) of the DDC system with existing or future facilities automation systems.

Chapter 5

Prequalification of DDC Systems and Vendor Selection

Introducing DDC Systems

Building Automation Systems (BAS) and the entire "controls and automation" industry are closely associated with high technology and the computer industry. Hi-tech industries, by their nature, undergo rapid changes. Constant evolution, and the highly competitive market, result in never-ending changes to systems hardware and software. The rapid changes affect the DDC vendors organizations (and in some cases, their ownership) in product marketing and systems support.

DDC manufacturers introduce new products, on average, every five years, and upgrade products annually. New, off-the-shelf DDC systems are introduced by major controls vendors who have long-term HVAC controls experience, and by "newer" DDC vendors with computer and systems expertise.

Building owners, the end users of DDC systems, benefit from the competitive situation because it offers competitive pricing. At the same time, they feel the pressure that comes from the frequent introduction of new systems, and the rise and fall of the very companies bringing them on the market. The owners' vulnerability is increased by their need for long-term systems support.

Many centralized BAS were upgraded to distributed DDC systems over the past decade. Distributed DDC systems became synonymous with HVAC control systems providing environmental control in the buildings. DDC systems provide reliable hardware platforms, sophisticated application programs, and are "user friendly" for operators and maintenance personnel. DDC systems have completely changed the controls industry, from systems development through implementation, support, and maintenance. Long-term performance of installed DDC systems depends on three components:

1. Selected DDC system(s)
2. System implementation
3. System support

DDC Systems Selection

Selection of DDC systems for individual projects should be in compliance with the long-term goals of the facility. Systems selected based on the lowest initial cost for every job could become a challenge for owners attempting to integrate incompatible systems or standardize certain systems and networks. The long-term view for DDC system selection is important, considering the cost of an installed system and the life span of installed automation systems, which is 10 to 15 years or longer. In making a long-term commitment (and a substantial investment), owners should protect their investment by selecting systems in accordance with the long-term automation goals of the facility.

The Bidding Process

Most DDC systems for building applications are selected in a bidding process. Bidding is widely used in the building industry despite the complexity of DDC systems (hardware, applications software, interoperability), systems performance, and the need for long-term systems support. In the bidding process, the initial cost of one piece of equipment (or service) is compared to another one with the same characteristics.

The bidding process does not take under consideration:

- Characteristics and intricacies of individual DDC systems (especially characteristics of their software features)
- The varying ability of different systems engineers to understand HVAC systems operations, and to write applications software that provides a fully automated environmental control system
- The future automation systems development goals for the facility
- The complexity of issues related to systems interoperability and networking
- DDC systems operations, management, and maintenance
- Quality of hardware, software, and communications options
- DDC vendors' long-term viability and provided system support.

Due to the complexity of DDC systems, the team involved in selecting a system must have a good understanding of DDC systems, their features, and their communications options. Assuming that many DDC systems will continue to be selected in bids on individual jobs rather than systems, the project managers and commissioning agents should focus on the following key items:

- System architecture
- Standalone controllers
- Application software
- System performance

DDC System Architecture

Selection of the DDC system's architecture (hardware, software and communications) should be with long-term facilities automation goals in mind. Selection of the most suitable architecture is probably the best protection of the owner's investment. Properly chosen system architecture assures future expansion of DDC systems, their upgrades, interfaces to third party systems and software, and interoperability of DDC systems over facilities' networks.

Generally, owners should be looking for "popular" elements in systems architecture:

1. In software programs, for example, the DOS operating system, Windows, and Lotus became so popular on the PC market, they became standards on their own merits. Among specialty software items, look for dBase for data base management, AutoCad, or AutoSketch for graphics software, Dynamic Data Exchange (DDE) for sharing information among individual programs, and so on. Many automation vendors base their automation software on these programs, or are using them in their systems.

2. Owners should select industry standard hardware such as IBM or Compaq PCs; Intel or Motorola processors; disk, tape, or CD drives for external data storage media; and RS-232, RS-485, Ethernet, or Arcnet communication modes.

3. Layered systems architecture, based on levels of standalone processing (i.e., on building, system, and equipment levels) provides an economical solution to hardware and wiring cost optimization, systems compartmentalization, and a base for future expansion.

4. Systems connectivity or interoperability is becoming more and more of an issue for facilities automation. Many BAS manufacturers changed their systems architecture (and their attitudes) from proprietary to "open" systems. Having an open system means that DDC vendors make their interface protocols available to other systems vendors, who can write interface protocols to these systems, if they desire to do so.

5. The open approach also applies to communication drivers. The popularity of open systems and widely accepted communication drivers is a positive sign leading to a broader base of future support of systems by vendors and systems houses. Many open systems and drivers fade away over time, but some become "industry standards," used and supported by many systems vendors, independent systems houses, and systems integrators.

6. There have been numerous attempts to standardize communications protocols for the automation industry. In the past few years, two standard protocols were approved by related agencies, and introduced for building automation systems in North America: the Canadian Automated Building (CAB) protocol, approved and released for public use by Public Works Canada in 1992; and BACNet, the standard communications protocol for BAS, approved and released by ASHRAE in 1996. Their success will be determined by their popularity among DDC vendors, and their acceptance by the users.

DDC Systems Hardware Architecture

Trends in the hardware architecture of automation systems are toward distributed processing. The most popular architectures are using controllers dedicated to controlling a single piece of equipment (e.g., a VAV box, or fan coil unit), larger, more complex mechanical systems (e.g., air handling units, roof top units), or a concentration of such equipment (e.g., equipment in a building or power plant). Selecting the right hardware architecture for the site is important for (a) future expandability of the DDC system, and (b) for cost optimization for the controllers, installation, and wiring.

DDC Systems Software Architecture

Software architecture follows the choices made for the hardware. Because it would not be economical (or necessary) to provide full software capabilities at each level of the system's architecture, full understanding of software features at each level of systems architecture (in individual controllers) is important. Allocation of software features at each level determines (a) the system's distributed processing capabilities, (b) standalone controls capabilities, and (c) the system's end-to-end reliability in case of communications failures.

DDC Network Architecture

Network architecture should be in compliance with the overall systems architecture. The network topography depends on the volume and speed of transmitted data, the communications protocols used, and on site conditions. For example, communications among application-specific or unitary controllers, controlling individual equipment, with a low data exchange rate, can utilize RS-232 or RS-485 lower-layer communication. Implement high volume, high speed data communication (e.g., via Ethernet or Arcnet) for data exchange on the site facilities network.

Standard methods of information access are equally important. Many facilities' networks standardize and support, for example, standard protocol packet methods such as Transmission Control Protocol/Internet Protocol (TCP/IP) information access, utilizing multiple networking software, such as NetWare (from Novell), LAN Manager (from International Business Machines), or DECnet (from Digital Equipment Corporation).

There are numerous options for systems architecture available from DDC vendors. The owner or systems integrator should design the most suitable option for the facility. However, availability of options also means they can be mismatched. For example, interoperability of an IBM-compatible computer with a Macintosh computer requires extra software and additional expense. Similar interoperability problems could be encountered with interconnecting DOS and OS/2 operating systems, and/or Ethernet lower layer protocols with an Arcnet protocol.

Network integrity and security is an important consideration for interoperability of real time systems on facilities networks. Secured access to real time systems is of utmost importance to prevent unintentional commands reaching large equipment (e.g., chillers, air handling units, etc.) or corrupting their control programs. Unintentional commands from unqualified operators to such equipment can injure anyone working on the equipment, or damage the expensive systems.

DDC Controllers

While selection of DDC systems architecture is for the entire facility, selection of DDC controllers is related to individual projects. Distributed processing allows selection of controllers that have a point capacity that matches the related application. Controllers should be located in the highest concentration of points, thus reducing wiring costs. Functions of the application software resident in individual controllers are also important in selecting DDC controllers. Communications capabilities and expandability of individual DDC controllers are equally important in evaluating the feasibility of controllers for the given application.

DDC Application Software

DDC application software is the brain in environmental control systems. Among the basic DDC features are:

Point Definition

Each piece of DDC application software has variations in defining points. Attributes of each point, and ease of their access for definition and on-line modification are important features that affect future DDC software maintenance.

On-line definition and modification of point attributes is important for systems start-up and future operations.

DDC Alarm Reporting

Alarm reporting is one of the most important features of DDC systems. Evaluate the number of analog alarm limits per point and their differentials (Hi/Lo, HHi/LLo, +/− of the defined values), their return to normal, binary alarms, scan rates for alarm updates, associated alarm messages, alarm acknowledgement, alarm inhibition, alarm point lockout (for systems down for repair, etc.), and other alarm handling characteristic of the DDC systems.

Time Scheduling

Time scheduling of building equipment and systems on an hourly, daily, weekly, or monthly basis, is one of the most common functions offered by DDC systems. Look for ease in entering new schedules, creating, modifying and disabling existing ones, scheduling holidays, and providing temporary schedules for scheduled events. A feature of global schedule change for an equipment group (e.g., all heating zones and HVAC schedules in a building), might be important for some facilities.

Application Programming

DDC application programs use structured line code or graphic programming for programming the control and automation logic for individual controllers. Important criteria for DDC systems selection are ease in program modification; code compilation; program download, and upload to controllers; and dynamic data exchange with other programs.

Operator Interface

The operator interface represents graphic displays; alarm and data presentation; dynamic screen updates; presentation of points in alarm; off-line, unreliable states; and so on, on the OWS. Ease of moving from screen to screen, alarm acknowledgement, time schedule modification, set point modifications, control loop tuning from the graphic screens, and other features are important for systems operators. The selection team should request a sample of operator interfaces generated for previous jobs to evaluate the systems engineer's ability to generate user-friendly Person Machine Interfaces.

Operator Access

Operator access to DDC software is via the OWS, remote terminals, hand-held terminals, and so on. Systems password protection, levels of passwords related to individual points, geographical areas, group of points, systems functions, network access, and network nodes, are important for systems integrity. Disabling, locking out and enabling individual points, groups of points, functions, systems, and programs or associated messages, are important for safe systems operation.

Report Generation

The function of report generation includes compiling data into reports, spreadsheets, and graphs. The common software required is an important consideration. Reports should be compiled into common formats that can be used by the users. Systems management, as well as facilities management, use alarm or management reports to evaluate building and systems performance. Dynamic data transfer of actual analog data into spreadsheets and other common statistical media should be evaluated.

History Data

Most DDC systems have history data reporting features. The main characteristics of the features that should be evaluated are:

- Size of history files
- Data scan intervals for updates
- Dynamic data exchange
- Statistical data manipulation (access) by the operator
- Data compilation into reports, tables, and graphs
- Data export to other media, or over the network.

Systems Diagnostics

Systems diagnostics evaluate the self-diagnostics and reporting features of DDC systems. Reports on communications performance, the on-line/off-line status of controllers and network devices, and other systems-related reports are important for systems managers.

Energy Management

Many DDC systems have been implemented to automatically control building HVAC systems and to contain energy costs. Energy management programs were first developed in the seventies, and have improved through the years. Evaluate the functions of the most common energy management programs, which are:

- Night setback—lowers building temperatures and rate of ventilation
- Free cooling—uses outside air for cooling, when available
- Optimal start—optimizes equipment start time
- Demand limiting—sets limits to avoid peak charges for electricity
- Duty cycling—cycles controlled equipment
- Central plant optimization—optimizes operation of chillers and boilers

DDC System Performance

Performance of DDC systems depends on the systems architecture, its software features, number of connected points, communications, and other system-specific issues. The evaluation team can get a feel for systems performance from the provided literature, system demonstrations or site visits to other users of this system. The clues to look for are: dynamic screen update; scan rate of field points; system's response time to changes of state; response to alarm; features and responsiveness of PID loop controllers; automatic tuning capabilities (continuous or on-demand); time required for software upload/download; system re-start after power failure; access time to points, programs, and files; ease of moving from screen to screen; and ease of making modifications from the colorgraphic screens, such as changes of point definition, modification of set points; alarm limits; alarm messages; print control; and enable/disable points.

Whether for single building or multi-building facilities, always select DDC systems with long-term objectives in mind.

DDC System Implementation

Implementation of the DDC system is a process of customizing off-the-shelf DDC systems to meet site requirements and to control connected building (HVAC) systems within design parameters. The process includes:

- Systems engineering, an expertise which should fully complement the HVAC design intentions
- Applications software development to provide software for the fully automated operation of controlled building systems
- Installation of DDC hardware and software
- Testing and validation of the installed system, and end-to-end analog accuracies, tuning of PID loops, tuning and balancing the system to provide desired building parameters, verifying application programs, and verifying operator interfaces
- Training of operators and controls technicians
- Final systems turn-over to the O&M department and end users.

While DDC systems development is on a par with the rest of the hi-tech industry, there is a lot to be done in the areas of systems implementation and long-term systems support.

One of the measures that would improve the current situation is site standardization of pre-selected DDC systems. This standardization would include:

  a. Pre-selection of the most suitable DDC systems and communications interfaces
  b. Utilization of standard control and automation routines, standardization of PMIs on OWS, standardization of alarms reporting and reports generation, and standardization of controls hardware, its installation, and operations and maintenance methods
  c. Continuous systems support, by establishing long-term relationships with pre-selected controls vendors for support of their systems.

Standardization provides a base for system selection, control and automation routines, engineering practices, and DDC hardware. It also reduces operating costs, and increases quality of services provided by the DDC vendor and the owner's O&M departments. Without site requirements for standardization and long-term vendor's support, DDC vendors often bid "bare bones, low cost systems." This could result in under-utilization of automation capabilities and features of DDC systems, and inadequate systems engineering and control of building and HVAC systems. With selection of the lowest bidder for every job, a multi-building facility could end up with several incompatible DDC systems, providing little more than the single loop controllers of the past, but for a much higher cost.

The long-term impact of the lack of policy on DDC systems selection and implementation, and the lack of long-term vendors' support, in combination with the selection of the lowest cost systems for every project, is illustrated in the following case study.

Case Study

The state purchasing guidelines require competitive bidding for all new and renovation projects. The facility, encompassing dozens of buildings, has an ongoing construction program. The program includes new buildings, and renovation of existing buildings or associated mechanical systems. For the past several decades, facilities management relied on outside consulting and contracting services for project implementation. Consulting engineering firms were selected in the competitive bidding for the design of new buildings and major renovations. Individual construction management firms managed the bidding and the construction process for each project. The facilities management department was maintenance oriented (as is tradition), without in-house engineering expertise to develop and support the facilities' goals for automation. Each consulting firm used its own approach to HVAC and controls design, with its own set of specifications and design standards. Different brands of controls systems from a variety of manufacturers were selected in the bidding process, based on the lowest cost for every job. With the introduction of Energy Management Systems (EMS) in the mid 1970s, some HVAC projects specified that requirements for HVAC systems be centrally monitored by the EMS. The energy management features of EMS were related to programmed start and stop of AHUs, and resetting set points of local pneumatic controllers. Since every building was designed by a different engineering firm, the number of EMS points per air handling unit varied from two to fifteen points per unit. Over the years, more and more points were added to the EMS. Years and many additional points later, the number of points on EMS reached its capacity; the system got overloaded, and became a huge maintenance problem. By that time the EMS was nearly useless, due to (a) limited functionality, (b) increasing number of EMS system related problems, (c) lack of systems maintenance and field calibration, and (d) lack of continuous training of operators and maintenance personnel.

It was decided not to expand the EMS, but rather to request DDC controllers for all new and HVAC retrofit jobs. The same purchasing policies, approach to project management, and DDC acquisition practices continued unchanged. In a few years, there were several different DDC systems implemented by a number of contractors. The quality of systems engineering, applications software, and utilization of DDC systems features varied by individual contractor. Most DDC controllers were designed as if they would be pneumatic single loop controllers. The cause was:

a. Lack of directions from facilities management
b. Lack of standard site specifications
c. Lack of standardization in DDC systems, systems implementation, and vendor support
d. Lack of in-house systems engineering expertise
e. Lack of DDC training of O&M personnel, and in general, lack of commitment to efficient building operations, and effective facilities management

The installed DDC systems became nothing more than very expensive single loop controllers, duplicating (in many cases) the functions of local pneumatic controllers. DDC systems implementation did not result in:

- Expected energy savings

- More accurate controls

- On-line HVAC and building problems analysis

- Alarm reporting

- Management report generation

- History data trending

- Reduced maintenance cost

- Increased operating efficiency

- Other benefits expected from DDC systems.

Despite the variety of installed controls and automation systems (pneumatic controllers, EMS, and DDCs), none of them fulfilled the promises of the engineers and contractors. With the passage of time, upper management became aware of the rising costs of operations and maintenance. Both HVAC and controls systems operation and maintenance became expensive; energy consumption remained unchanged or became higher due to implementation of new HVAC systems in buildings. Manpower requirements and associated cost for environmental control systems operation and maintenance were growing with every installed system. The training cost for controls technicians was rising, yet the O&M personnel were not able to keep up with the variety of installed systems. Cost of service and maintenance contracts with different controls vendors and contractors increased with every new installed system. The situation cried out for a change.

Unfortunately, there are no "quick fixes" for a case like this one. To catch up with the rest of the world, this facility has to adopt major policy changes toward development of:

- Long-range plans (including facilities automation)

- Design standards

- Performance standards

- Standardization of material

- Energy conservation measures

- O&M procedures

- Procurement procedures

- Methods of budgeting, justifying, and funding environmental controls related projects

- Training in-house O&M personnel

- Long-term systems support from systems vendors

To upgrade existing controls systems (and in many cases associated HVAC systems), and to assure systems interoperability, requires major capital investment. Even with the availability of funding, considering the large number of fully operational buildings and HVAC systems in this case study facility, it could take several years to upgrade all existing systems.

For situations similar to the case study, facilities management has to adopt new approaches and should develop:

- An overall facilities automation master plan

- BAS master plan including interoperability

- Site standards

- Systems selection criteria

- Training criteria

- Funding and budgeting criteria

- Implementation time lines

- Evaluation criteria

Successful DDC systems implementation is based on defined facilities master plans, site specifications, defined requirements for engineering, and software development. These are prerequisites for successful systems implementation, systems operation, and long-term systems support.

DDC Vendor Support

DDC vendor support is important for systems upgrades, training, service, and maintenance. Very few sites, if any, can exist without some form of vendor support. In the past, support was limited to service and repair of the installed hardware. That is rapidly changing because of the complexity of computers, and controls operating software, and because of the ongoing need for systems upgrades. DDC systems support is important for the design and implementation phases of the project, as well as follow-up support.

Most DDC systems in the United States are:

a. Developed, produced, marketed, sold, engineered, installed, and supported by their manufacturers or vendors

b. Developed, produced, and marketed by a DDC vendor; sold, engineered, installed, and supported by others, such as dealers or manufacturer representatives

c. A combination of (a) and (b).

Large, multinational controls companies have established local branch offices to cover their customer base. Local offices are staffed to provide full support for their customers. These branch offices are fully dedicated to sales and support of their own product lines in the respective geographical areas. The home office is responsible for product development, manufacturing, marketing, sales, and technical support of the branch offices, as well as training of the staff and their customers. The home office and the branch share legal responsibilities and contractual obligations to the end users for products and services rendered.

Another method used by DDC vendors to market and sell their products is through independent local dealers, franchises, or contractors. The DDC vendors provide product development, manufacturing, marketing, and training of the network of independent dealers and contractors. The local dealers or contractors buy the systems from the DDC vendor, but most DDC vendors have no financial obligations to support the dealers or contractors. End users deal with the local authorized dealers or contractors, and have no legal connection or contractual obligations with the DDC vendor.

Independent dealers or contractors may carry one vendor's product line, or different products of non-competing vendors. Some dealers or contractors carry only the DDC system complementary to their main business interests, such as mechanical or HVAC contracting. Many DDC dealers and contractors become loyal to their vendors and display continuous commitment to support the systems they sell. Probably, equal numbers of dealers or contractors have changed product lines in response to market demands and/or their own financial needs. Business interests of independent dealers and contractors and the local market conditions may cause them to abandon existing product lines and pick up other, more profitable products. As a result, some owners end up with discontinued products, failed vendors' systems, or systems which cannot migrate to a newer generation at the end of their useful life cycle.

Either of the marketing and sales styles can provide adequate systems support to the owner/end user at any given time. However, owners interested in a long-term contractual relationship, or partnership for support of DDC systems, must consider checking the business style of the vendor. Background checking of involved companies during the selection process is not only prudent, but essential.

Objectives for DDC Selection

The selection team must set goals before starting DDC selection. Two recommended goals are:

1. To pre-select DDC systems that comply with site long-range development objectives for facilities automation, operations, and maintenance. Pre-selected systems should also comply with site design criteria, standards, and site interoperability requirements. The site objectives should clearly state site-specific requirements for DDC systems, implementation, and systems support.

2. To pre-select DDC vendors who can provide DDC systems in compliance with the facilities' long-range goals; have a proven track record for long-term DDC systems development and migration, and development and support of standard or open protocols for third party interfaces; have a track record for long-term systems and customer support; and have local representation with adequate engineering, installation, and service support capacity.

By pre-selecting certain DDC systems, the owner is establishing a long-term relationship with the pre-selected DDC vendors, or their local representation. In other words, in the selection process consider the vendor as well as the system because when you pre-select one, you're automatically pre-selecting the other.

Evaluation Team

To create an evaluation team, the owner's commissioning agent (facilities engineers, or other in-house entity responsible for long-term development of facilities or building automation systems) should assemble a number of experts to pre-qualify DDC systems and vendors. The team should include:

- Facilities systems engineer or systems integrator
- Controls technician or systems operator
- Operations and maintenance management
- Financial officer

If the facilities are lacking some of this in-house expertise, hire qualified consultants to substitute for the suggested expertise. Nevertheless, facilities management should be involved and have a leading role in the process.

An appointed team leader should report to facilities management. A professional approach in dealing with DDC vendors and contractors throughout the selection process is important. After all, they are dedicating their time and effort to the process with a distant possibility that they will be one of the selected vendors for the site.

DDC Selection Requirements

The defined requirements must be consistent with site-specific conditions, standards, and specifications. They should be published in the form of clear statements to assure uniform interpretation by all participants.

Preselection Requirements for DDC Systems

The DDC systems should meet the following requirements:

1. Pre-selected DDC systems shall comply with the requirements specified in the site performance and material specification.

2. To be considered for the site, DDC systems shall have been in production for over one year, and installed, and operational on job sites similar to the owner's site for at least one year.

3. All proposed communications and interface protocols shall have been operational on at least one job site similar to the owner's site, for the minimum duration of one year. In absence of such a site, the DDC vendor shall set up communications testing at the owner's site free-of-charge for a minimum of two weeks. The DDC communication shall be tested on the longest available communication wiring on the site.

4. The DDC systems architecture shall comply with site conditions.

5. Operator's interfaces shall meet site-specific requirements.

Preselection Requirements for DDC Vendors and/or Contractors

DDC vendors and/or contractors should meet the following requirements and/or disclose the following information:

1. DDC vendors shall demonstrate their commitment to long-term systems development by demonstrating past systems development and their migration paths.

2. DDC vendors shall demonstrate their commitment to open systems architecture by demonstrating adherence to major communications standards, industry, and open protocols and drivers. These items are to protect the owner's investment for the selected automation systems, as well as to assure optimum life-cycle cost for the installed systems.

3. DDC vendors shall demonstrate commitment to systems and customer support from their headquarters as well as from the local office.

4. DDC vendors shall maintain a local office with sales, systems engineering, installation, and service support.

5. Independent franchises of Building Automation Systems shall disclose what other products or services are carried or provided by the office, and which products or business interests are the backbone of their business.

6. DDC vendors or contractors shall provide the owner with a statement of their financial situation, annual sales and service contracts, and other requested financial information.

7. DDC vendors and contractors shall provide business references related to their customer base, suppliers, and others as requested by the owner.

8. DDC vendors and contractors shall provide the owner with the number and qualifications of their engineers, technicians, and support personnel in the local office, and their capabilities to respond to customer needs and emergency situations.

9. DCC vendors shall disclose how long the proposed product line (DDC system) has been sold and supported by the local office. What other controls and DDC systems were sold and supported by the office prior to the current product line?

10. The main office or the DDC manufacturer shall make a commitment to technical support and training of the end-user.

The Preselection Process

The following steps are to aid engineers and commissioning agents in conducting a DDC pre-selection process:

Step 1: Request for DDC Documentation
Step 2: Development and Evaluation of a Pre-Qualification Questionnaire
Step 3: DDC Systems Demonstration
Step 4: Site Visits
Step 5: Visit to DDC Vendor's or Contractor's Home Office
Step 6: Final Evaluation and Ranking

Step 1: Request for DDC Documentation

Mail a formal request to DDC vendors and/or contractors for sales literature and other systems documentation. The letter should be specific in its purpose and requested information. Depending on the expertise of the owner's team, the letter should state what kind of documentation is requested. Usual color brochures are meant to attract customer attention; however, they provide very little engineering information. Most vendors will provide more detailed software and hardware documentation if requested by the potential customer.

Some DDC vendors or contractors will offer a personal visit and systems demonstration following the request. It is important for the team leader to control the situation, and decline a visit if the team is not yet ready for a demonstration. The entire process is a learning experience for the team members who need adequate time to digest the provided information. Upon receipt of information, the team members should review the documentation, evaluate, and rank the individual DDC systems.

Step 2: Development and Evaluation of a Pre-Qualification Questionnaire

The purpose of a pre-qualification questionnaire is to obtain basic information about DDC systems, vendors, or contractors. Send the questionnaire to DDC vendors and contractors active in the area (and pre-selected in Step 1). Team members should develop the questionnaire jointly to cover the wide spectrum of needs and interests.

Following is an example of a prequalification questionnaire.

Prequalification Questionnaire

A. Vendor Related

1. DDC vendor has local representation within 50 miles of the site   Yes__ No__
2. Are other non-DDC products carried, or services provided, by the local office? (Name them and indicate % of revenues for each item on an attached list)   Yes__ No__
3. Provide Annual Report of the DDC vendor and the local representative   Yes__ No__
4. Provide resumes of engineers, consultants, project managers, and other personnel associated with the project   Yes__ No__
5. Number of systems engineers in the local office ____
6. Number of DDC technicians in the local office.____
7. Electrical installation provided by local office?   Yes__ No__
   Directly____ or Subcontracted____
8. Mechanical installation provided by local office?   Yes__ No__
   Directly____ or Subcontracted____
9. Number of electricians in the local office.____
10. Number of mechanical installers in the local office.____
11. Number of technicians responding to service calls.____
12. Emergency response 24 hrs/year round/within 4 hrs   Yes__ No__
13. Number of technicians responding to emergency calls.____
14. Number of years DDC system carried and supported by the local office: **less than 1____, more than 3____, more than 5____**
15. Customer training is provided by: **local office____ main office____**
16. Provide three references (with names and telephone numbers) of similar sites with DDCs installed over the last three years:

    _____
    _____
    _____

17. Vendor will submit DDC systems literature as requested   Yes__ No__
18. Provide on-site systems demonstration.   Yes__ No__

B. Long-Term Support

1. Vendor requires DDC software licenses for the system   Yes__ No__
2. Site license available   Yes__ No__
3. Software upgrades will be available to the owner **with____/without____** service agreement at **cost____/no cost____** upon release
4. Spare parts available from **(company/location)** _____
5. Training available from **(company/location)** _____
6. Service agreement available from **(company/location)** _____
7. Application software support available from **E-mail____ Internet____ on CD Rom____ in person____ via telephone____**
8. ____ **year** warranty for DDC **software____/hardware____/ application programs____/installation____/field hardware____** provided with the standard contract
9. Extended warranty available   Yes__ No__
10. Partnership agreement available for **extended warranty____/ systems support____/training____/engineering____/pricing____**
11. Provide samples of contracts, including partnership agreement.   Yes__ No__

C. Implementation

1. Systems engineering support available from the local office for HVAC design   Yes__ No__
2. DDC systems engineering by the local office   Yes__ No__

3. Controls design available from the local office   **Yes__ No__**
4. Application programming available from the local office   **Yes__ No__**
5. Engineering documentation provided by the local office   **Yes__ No__**
6. System testing and validation by the local office   **Yes__ No__**
7. Training available from the main office   **Yes__ No__**
    local office   **Yes__ No__**
8. Test operation provided for one year   **Yes__ No__**
9. As-built documentation including operation and maintenance manuals   **Yes__ No__**

D. Communications and interfaces

1. Communication methods, media, and speed in the building _____
_____

2. Communication methods, media, and speed between buildings _____
_____

3. Third party controller interfaces _____
_____
_____
_____

4. Common communications protocols _____
_____
_____
_____

5. Willing to set up a 14 day communication test at the owner's site on the existing media, utilizing standard DDC components and software, at no cost to the owner.   **Yes__ No__**
6. Provide a list of implemented interfaces, protocols and drivers (attach list) **Yes__ No__**
7. Provide a list of customer sites where similar interfaces are implemented. (attach list)   **Yes__ No__**

E. Provide the following systems-related descriptions or engineering and software data sheets

1. DDC systems architecture
2. Systems security
3. Operating Work Stations and Person-Machine Interface
4. Point control/set point adjusting from the OWS
5. Alarm reporting
6. History data format and its graphical trending
7. Report generation
8. PID self-tuning from OWS
9. Interfaces to other systems
10. Data transfer to other media and systems
11. "Standard" applications programs designed by your office or provide programming samples for:
    a. Free cooling
    b. Discharge air temperature and volumetric control of a VAV system
    c. PID of a hot water reheat coil valve
    d. Zone valve control of a steam heating system
12. Submit a sample of:
    a. Controls drawing
    b. Point summary (I/O list)
    c. Application programming
    d. User's manual
    e. PM schedule of your system
13. Specification of OWS, including communications hardware and software.

Step 3: DDC Systems Demonstration

Upon review of the systems documentation and the returned questionnaires, the team should be ready for DDC system demonstrations. Demonstrations can be set up at the owner's site or at the DDC vendor's or contractor's office.

A few words of caution about customer demonstrations:

1. They are carefully orchestrated and rehearsed by sales or marketing professionals whose sole role with the company is to demonstrate and sell systems.

2. They all claim to be customer oriented. In fact, very few demonstrations are customer oriented. This is because "systems demonstrators," in most cases, have limited or no real life systems experience with DDC systems applications, operations or maintenance. Therefore, even the best demonstrations that are tailored to the customer needs fall short of addressing real needs of specific applications.

3. The purpose of systems demonstrations is to show all the systems features to the potential customer in a controlled, pleasant environment.

4. Customers should listen to the presentation, and ask questions pertaining to specific applications or site-specific conditions. Customers should always expect answers. If not during the demonstration, then shortly afterwards in a follow-up call or correspondence.

5. Use systems demonstrations to learn more about the company, its management philosophy, commitment to customers, support, financial standings, and other important aspects. Remember, you are establishing a long-term relationship with a company that should provide DDC systems implementation as well as long-term support of the systems.

The owner's team should follow up with an evaluation of systems demonstrations. (DDC vendors are also evaluating their potential customers). Evaluation of systems demonstrations is subjective. Each member of the team may have a different degree of favorable or unfavorable impressions, depending on the demonstration and his or her personal experience. Team members should focus on evaluating each system in relation to site-specific conditions, and from their own point of expertise.

Step 4: Site Visits

Up to this point, the controls vendor or contractor has had control over the presentation of their DDC system. The owner's team has learned about the DDC system and the company only from the vendor's representatives. The team is now ready to compare the presentations with real operating experiences at DDC sites similar to the owner's site. Site visits are the most important part of the pre-selection process. Conversations with facilities engineers, operators, and technicians can shed more light on the system than all the literature provided by the DDC vendor. The team should learn from the other personnel about their experience with the installed DDC system. Site visits can be set up either by the DDC vendor or directly by the team leader. In either case, it should not be another vendor demonstration, but rather an informal exchange of ideas and experiences between two groups of end users.

Step 5: Visit to DDC Vendor's or Contractor's Home Office

The visit to a home office can be combined with a DDC systems presentation, or scheduled separately. The team members should have a chance to be exposed to the "corporate culture" of the home office. This is an opportunity to find out about the company's commitment for long-term system support, development plans, and their relationship to the local office. Another important reason for home office visits is to get an understanding of the structure of the vendor's business, the financial standing of the company, and their plans for future systems development and engineering support.

(I fondly recall a case when the local office promised an open architecture; the regional manager provided a corporate jet to fly the customers to his home office. When the question of open architecture was raised, the corporate answer was NO, despite the vendor's disqualification from bidding on the job.)

In a sales situation, the local office can make promises (e.g., related to future systems, interfaces, protocols, etc.) that cannot be kept by the home office. On the other hand, the home office can make commitments to the owner (e.g., related to evaluating the owner's networking needs, or setting up national accounts, open book pricing, partnership agreements, etc.) that are beyond the authority of the local office. These possibilities depend on the business orientation and relationship of the participants. The business goals of a local office are to respond to the daily needs of their customers; the business goals of a home office are related to long-term product development, continuous systems support, and long-term market development.

Step 6: Final Evaluation and Ranking

In the final phase of the pre-selection process, the team concludes its work by evaluating each DDC system individually. The following are examples of evaluation checklists. The checklists should be modified by site-specific conditions and requirements. When evaluation by individual team members is complete, the team leader or commissioning agent arranges a meeting for the final evaluation and ranking of pre-selected DDC systems and vendors. (See page 128 for a sample ranking checklist.)

Checklist for DDC Selection and Ranking: System Architecture

(Rank 10 = Highest/1 = Lowest)

Item	Vendor			Remarks
	A	B	C	
Architecture				
System architecture				
Software architecture				
Distributed processing				
Open protocol				
Third party drivers				
Industry standard protocols				
Communications				
Communication methods in buildings				

Checklist for DDC Selection and Ranking: System Architecture—contd.

(Rank 10 = Highest/1 = Lowest)

Item	Vendor			Remarks
	A	B	C	
Proposed protocol				
Communications speed				
Communications media				
Communication methods between buildings				
Proposed protocol				
Communication speed				
Communications media				
Interfaces to site LAN				
Proposed LAN nodes				
DDC access from LAN nodes				
LAN security/BAS security				
Communications software option				
BAS dial-up options				
Interface to maintenance management system				
Distributed Controllers				
Building controllers				
Application logic in building controllers				
Data base capacity in building controllers				
Application specific controllers				
Operating logic in application specific controllers				
Data base capacity of application specific controllers				
Third party controllers				
Operating logic in third party controllers				
Data base capacity of third party controllers				
OWS and Operator Interfaces				
Operator interfaces				
Local/building interfaces				
Hand held (maintenance) terminals				
OWS data base capacity				
OWS common (popular) features				
processors				
hardware				
external storage				
dBase				
Excel				
Word Processing				
Spread sheets				
AutoCad				
Graphic programs				
DDE				
Access				
Point access				
Database access				
Database up/down load				
Applications program access				
BAS access and security				

Checklist for DDC Selection and Ranking: Front End OWS

(Rank 10 = Highest/1 = Lowest)

Item	Vendor			Remarks
	A	B	C	
OWS Hardware				
Computer(s)				
Printer(s)				
Communication cards/modems				
CRT(s)				
External date storage media				
Surge protection				
Operator Interface				
Number of color graphic screens				
Number of points per color graphics				
Structure of PMI				
Quality of color graphic display				
Screen update time				
Point binding				
Point control/modification from color graphic display				
Change of state/point status on color graphic display				
Log on/log off procedures				
Database download/upload procedures				
Database access from OWS				
System reboot procedures				
Data archiving procedures				
Report Generation				
Standard reports				
All point reports				
Out-of-service report				
Point locked-out/down reports				
Analog limit summary				
Analog set-point summary				
Equipment run-time report				
System status report				
Alarm Reporting				
Alarm display				
Alarm summary report				
Alarm acknowledgment				
Alarm messages				
History Data, Archive				
Size of history files				
Number of trend graphs				
Points per graph				
Time intervals per graph				
Statistical tools				
Export to Other OWS				
Data export				
Graphic export				
Dynamic Data Exchange				
Spread sheet export				

Checklist for DDC Selection and Ranking: Building (or Application-Specific) Controller

(Rank 10 = Highest/1 = Lowest)

Item	Vendor			Remarks
	A	B	C	
Controller				
AC power supply				
Battery back-up				
Wire termination				
Spike/transient protection				
Grounding				
Access door/locks				
Local access				
Local display				
Controller's modularity				
expendability				
configurability				
redundancy				
External transformers				
power sources				
I/O relays				
CPU				
CPU make/model				
Memory capacity				
Communication port				
Network interfaces				
Standard bus				
Surge/lightning protection				
I/O Modules				
I/O modules capacity				
A/D conversion				
No. of analog input modules and expandability				
No. of analog output modules and expandability				
Output Voltage/amperage/ohms				
Input Voltage/amperage/ohms				
Pneumatic outputs				
Signal conditioning				
Analog filters				
No. of digital input modules and expandability				
No. of digital output modules and expandability				
Output Voltage/amperage				
Input Voltage/amperage				
Signal conditioning				

Evaluation Checklist for DDC System Application Software

(Rank 10 = Highest/1 = Lowest)

Item	Vendor			Remarks
	A	B	C	
Definitions				
Point definition				
Point modification				
Alarm definition				
Alarm limits definition				
Alarm message definition				
Alarm modification				
Alarm routing				
Alarm log				
Display of alarms on associated graphics screens				
Time schedule definition				
Schedule modification				
Temporary occupancy schedules				
Schedule overrides				
Set point definition				
Set point modification				
Application Programming				
PID loop definition				
PID loop modification				
PID tuning				
Math routines definition				
Math routines modification				
Ease of programming				
Line programming				
Graphic programming				
Programming language				
Program modification				
PMI programming				
Screen modifications				
History report programming				
History files				
File export/import				
Trend graph programming				
Report generation				
Energy management programs				
Demand limiting				
Duty cycling				
Optimum start time				
Nite set-back				
Free cooling (economizer)				
Plant optimization programs				
Support of intelligent sensors				
Support of intelligent actuators				
Analog point tuning on-demand				
Continuous tuning of analog I/Os				
PMI color graphics				
Tree design of color graphic				

Evaluation Checklist for DDC System Application Software—continued

(Rank 10 = Highest/1 = Lowest)

Item	Vendor			Remarks
	A	B	C	
Rapid screen updates				
Point commands from the color graphics				
Software and Database				
Data sharing				
Object oriented SW				
Modular SW				
3rd party SWs				
Interfaces to popular SWs				
Auto configuration of database				
Object searching				
Global replacement				
Distributed SW				
SW access/security				

Evaluation Checklist for DDC Vendor/Contractor

(Rank 10 = Highest/1 = Lowest)

Item	Vendor			Remarks
	A	B	C	
Home Office				
Financial profile				
New product development				
DDC system migration				
DDC system support				
Customer support				
Contact person(s)				
Training facilities/literature				
Local Office				
Number of employees				
Number of engineers				
Number of technicians				
Can install controls				
Can provide electrical work				
Can provide mechanical work				
Other services				
Application programming experience for HVAC controls				
Application programming experience for power plants				
Application programming experience with communications				
Application programming experience with interface protocols				

Evaluation Checklist for DDC Vendor/Contractor—continued

Item	Vendor			Remarks
	A	B	C	
Provide 24 hr. support				
Provide hardware repair				
Provide spare parts				
Provide HVAC systems troubleshooting				
Provide DDC engineering support				
Provide HVAC/energy audits				
Provide performance contracting				
Provide HVAC & DDC design/build jobs				
Provide energy contracting				
Documentation support				
Engineering support				
Local references				
Local site visits				

Final Ranking of DDC Systems, Vendors, and Contractors

The evaluation process, along with site specifications, can give a team a degree of confidence to rank and select DDC systems suitable for site-specific conditions. Selecting a DDC system begins a long-term relationship with the pre-selected DDC vendor and/or contractor. It is also the start of standardization on pre-selected systems, training of O&M personnel, and a standardization of operating and maintenance procedures.

Note that it is mutually advantageous to develop personal contacts at the upper management level between the facilities and the pre-selected vendors, to demonstrate intentions for long-term relationships.

Final Ranking Checklist of DDC Systems Vendors/Contractors

By Department Members	Vendor			Remarks
	A	B	C	
Engineering				
Operators				
Maintenance				
Financial				

Chapter 6

DDC Systems Engineering and Application Software Development

Systems Engineering Activities

Perhaps the most important engineering activity during the design/development phase of the project, with the greatest impact on the performance of the environmental control system, is the DDC systems engineering and generating of application software. This phase is the least understood (and appreciated) outside of the controls systems engineering community. However, DDC systems engineering and related application software (such as controls logic, automation programs, report generation, Person Machine Interface, communications, and other relevant programs) determine:

    a. The future operation of the environmental controls system, and consequently the quality of the building environment

    b. The level (degree) of automation for the facility

    c. Energy savings for the controlled building(s)

    d. Reduction of operating and maintenance costs due to

- Designed hardware and application software
- On-line detection of related environmental control systems and building problems
- On-line analysis of problems, and assignment of appropriate trades
- On-line transfer of information to a job scheduling or maintenance management system
- Generation of energy, problem (alarm), and management reports
- Reduced number of nuisance calls
- Generation of trends and history data for situation analysis

    e. Reporting of fire, safety, and security alarms from the connected buildings

    f. The volume (and value) of information exchange with other facilities automation systems

Many traditional project teams are still hardware oriented, and have limited understanding and appreciation for the discipline so important for the operation of modern facilities. This may have its roots in years of specifying single loop pneumatic controllers for building HVAC controls. Some engineers on the design team perceive DDC systems as if they were "solid state" single loop pneumatic controllers. This perception prevents full utilization of the system's available features in facilities automation. The limited vision, along with the desire to select the lowest cost DDC system, have adverse effects on the design of the environmental controls system, its future performance, and on the degree of automation of the facility. Limited understanding most likely leads to underutilization, whereas full understanding of the DDC system's capabilities, and development of its applications software, provides true "building automation."

    DDC software development is a large and important part of building automation systems engineering. Systems are like icebergs with mechanical features visible above the water, and larger software and automation features hidden beneath the water. Underestimating the invisible features can lead to embarrassing engineering failures (remember the unsinkable Titanic!).

Systems engineering and the development of application software should start with an understanding of the building performance requirements, HVAC design, sequence of operation, and site-specific conditions. Early involvement of the application engineer is essential for configuring the DDC system's hardware and application software to meet the design and facilities automation requirements. Participation of a DDC systems engineer also influences and enhances the HVAC systems design, resulting in a much better environmental control system. Teamwork results in reduced costs to the owner due to optimum design of HVAC systems, and a customized DDC system for the designed application. This reason should be enough for project managers to bring systems engineers into the loop at the earliest possible stage of HVAC projects.

The importance of the participation of a DDC systems engineer in the process, and the resulting savings are demonstrated in the following situation:

A building renovation project was completed without the participation of a DDC systems engineer in the process. Due to a tight job schedule, the HVAC design went out for bid without proper review by the pre-selected DDC systems vendor. As in many cases, the DDC system was bid to mechanical contractors. Construction started prior to the project receiving the mechanical contractor's bid, which also included the DDC systems cost. The general contractor and the project manager had reviewed the proposal, and were alarmed by the high cost of the DDC system compared to the budgeted cost. After the initial surprise, and in paying attention to recommendations to cut controls, the project manager organized a working session with a facilities engineer, HVAC design engineer, and a controls systems engineer. First, they reviewed the original DDC scope of work on which the budget was based, and compared it to the bid's scope of work. They found that the original scope was expanded and the number of control points was increased from the original scope. Second, they reviewed the HVAC design from the facilities operations' point of view, made modifications, and complemented the HVAC design with a DDC systems design (field points and systems architecture). Their efforts resulted in: (a) modifications of the HVAC design, and (b) changes of the DDC scope of work.

The engineers' task was to come up with the best solution for the building even if it would result in modifications to the existing documentation. The review process took two sessions. Optimization of HVAC design and resulting changes of the DDC system's scope of work resulted in a 30% cost reduction to the DDC bid proposal. Another major cost reduction was achieved with modifications to the HVAC system.

Engineers working individually tend to "over-design" their systems in an effort to provide sufficient allowances and safety margins. To design "the best," they add features and components that compensate for possible shortcomings of other, associated, systems. Teamwork among facilities, HVAC, and systems engineers often results in optimum design since they have an intimate knowledge of their areas of expertise, which they apply to systems design.

Systems engineering is a complex process of transferring human ideas into application software and into electronic signals. Because there is no generic DDC system on the market, systems engineering is specific to a particular DDC system.

Essential Input for Systems Engineering

Essential input to the systems engineering process includes: an appropriate design of the HVAC system for the application, a clear definition of operating parameters, a description of the operating logic (Sequence of Operation), and a definition of site requirements for operations and maintenance. The systems engineer should receive the following items:

Documentation from the HVAC Engineer

1. HVAC and controls job specifications
2. HVAC systems schematics
3. HVAC and utilities distribution system layout in the building
4. Preliminary point list
5. Sequence of Operation
6. Specification of HVAC equipment with dedicated controllers
7. Building/systems control parameters
8. Definition of interfaces to other building, utilities and control systems
9. Any other job-specific issues and requirements

Site Requirements from the Owner

1. Site design criteria, including naming conventions
2. Site DDC specifications
3. Site requirements for interface to existing DDC systems and facilities networks
4. Preferred communications methods within buildings
5. Preferred site communications methods external to buildings
6. Requirements for operator access, to control and monitor environmental control systems locally and centrally
7. Requirements for systems security
8. Site standardization on controls hardware and software routines for control, operation, and maintenance
9. Maintenance practices and interaction with site Maintenance Management Systems (MMS)
10. Required energy management functions
11. Networking and information transfer requirements to and from other facilities' computerized systems on the same network
12. Any other site-specific issues and requirements in relation to the project

Job and Site Walk-Through

Site surveys and walk-throughs are essential for renovation and design/build projects. Walk-throughs are recommended even for projects where the consulting engineer has conducted a thorough engineering site survey.

For new construction jobs, and jobs designed by consulting engineering firms, the walk-through familiarizes the systems engineer with the new construction, site conditions, and O&M practices.

Review of HVAC Documentation and Finalizing DDC Scope of Work

The final DDC scope of work should be based on an engineering document that has been developed by the engineering team and approved by the review teams. Finalizing a DDC scope of work for new construction is more straightforward than for retrofit jobs. For new construction, the HVAC engineer designs and specifies the mechanical components to be controlled by the DDC system. All interfaces are defined at the time of design. The design team can decide on the degree of automation, define systems interoperability, and define operational and maintenance procedures. Because the entire design team works together during the same timeframe, the DDC scope of work is a natural outcome of the process.

In retrofit installations definition of the DDC scope of work is a challenge. In many retrofit jobs there are existing building systems and equipment that are outside of the immediate scope of work. Proper control of such systems, and/or monitoring of such equipment, is essential for control and monitoring of the entire building environment. Among such systems are: exhaust systems which have to be controlled to maintain designed air balance; perimeter radiation which has to be controlled and coordinated with the new air system, hydronic systems, pumps, and condensate return systems. Another category that can influence the final scope is unaccounted equipment. Such equipment can be anything from a non-functioning air handling unit to a hot water pump in the basement. Some renovations are for building systems only, and exclude existing systems such as dedicated fume hood exhaust systems. Even though not directly related to the project, they are part of the renovated space, and can influence the final performance of the renovated system.

The submitted HVAC documentation, and the site requirements along with an engineering walkthrough, should provide sufficient information for the engineers to prepare the final DDC scope of work.

DDC Systems Engineering Tasks

Because systems engineering is done by DDC vendors or controls contractors from outside of the traditional design teams, there is limited understanding by the design team of the tasks associated with engineering of DDC systems. A better understanding of systems engineering tasks, their sequence, and associated activities would help the design and project management process, especially when it comes to engineering changes. Costs associated with changes of hardware specifications are well understood by project managers. Costs associated with software changes are often a surprise. Better understanding of systems engineering activities and their sequence may save money for re-engineering or programming.

One of the common changes during systems engineering is the re-numbering of rooms in the designed building. Room numbering tends to change during design, and sometimes after design completion. Project managers are usually surprised at the extent of software modifications required to change the room numbers of a large building. But in a VAV controlled building with boxes in every room there is a temperature sensor, damper actuator, maybe a reheat coil valve actuator, and a flow sensor, bearing the room number in their designation. The acronyms associated with room 101 may look like this: VAV101, R101T, VAV101DMPR, VAV101RHC, and VAV101FLOW. Each acronym then appears in several locations in the application software: in point definitions, alarm definition, set point definition, and in all software features and associated programming blocks. In the absence of a "global change," the programmer has to access every location of that acronym in the software, and make the required change—a time consuming task for a building with several hundred rooms. The charges for the number of programming hours it requires to do a "rewrite" for a seemingly simple change can be astonishing. The cost can be avoided if the change is requested prior to the start of writing the application software.

The following is a list of the major systems engineering tasks:

1. Engineering the DDC system architecture
2. Design of communications methods
3. Controls application engineering
4. Specification of DDC controllers
5. Point definition for each controller
6. Instrument specifications, and instrument schedules
7. Definition of DDC operator interfaces
8. Locating field gear and controllers throughout the building
9. Electrical design
10. Ordering materials (valves, dampers, air flow stations, actuators, etc.)
11. Development of operating logic
12. Definition of points, limits, set points, alarms, etc.
13. Definition of energy management functions
14. Application software review and approval
15. Application programming for each DDC controller and front end OWS
16. Development of colorgraphics for OWS
17. Development of report generation and alarm reporting
18. Application software testing
19. Software download/upload to controllers and front end
20. Testing of communication methods (if required by owner)
21. Review and approval of operator interfaces
22. Documentation review
23. Vendor (factory) testing of hardware, software, and interfaces, prior to shipment
24. (Factory) acceptance testing by the owner, if required
25. Ready for installation

Commissioning/TQM Process for DDC Systems Engineering

Objectives

The main objective during systems engineering is to engineer a DDC system, including field gear, hardware, application software, and communications, that controls the building environment at design parameters. To achieve this objective the commissioning agent must focus on the following:

1. Systems engineering to meet site design standards and criteria, such as site performance specifications, standardization on field hardware as per the material specification, methods of communications as requested by the facilities (and tested at the owner's site), standardization on documentation as requested by the facilities, and other relevant site specific requirements.

2. Systems engineering to meet job specifications as per the provided job documentation.

3. Development of application software based on the reviewed and approved Sequence of Operation.

4. Systems engineering to provide full automation for the connected HVAC systems and the controlled building.

5. Systems engineering to meet site-specific requirements and conventions for systems operations and maintenance practices.

The Systems Engineering Team

Teamwork among systems engineers, the HVAC design engineer, and the facility engineer or O&M personnel is important for DDC engineering and for the development of software that fully meets the defined needs of the job and facility requirements. Though considered a development phase, in fact, systems engineering bridges over to the implementation phase of the project. During applications engineering the applications software is not only developed, but also tested and loaded into the DDC controllers and Operators Work Stations. The only outstanding software-related items should be minor software modifications such as tuning parameters, set points, alarm limits, and the binding of real points (their readings) with their display fields on the colorgraphic operating screens. These items are usually done during the installation and commissioning phases of the implementation process.

Project managers and commissioning agents should request applications software development, testing, approval, and their loading into the controllers and OWS at the vendor's premises. If the applications software, communications interfaces, and operators' PMIs are not finalized and operational on the vendor's test bench (and witnessed by the owner's commissioning agent) at this phase of the project, the project team can expect great difficulties and delays during commissioning and start-up of the system.

At present, most systems engineering work is done behind the closed doors of control vendors or contractors, with limited participation by the owner's and HVAC engineers. Such practices may lead to many disappointments between the controls vendor on one side and the HVAC and owner's engineers on the other. The cause for disappointment could range from minor discrepancies between the HVAC design and the DDC system, to major differences between the functionality of the delivered DDC system and the owner's and HVAC engineers' expectations. A typical defense by control vendors or contractors when these discrepancies are brought up during systems operation, is "but that's not what we were told.", "That's not how I understood it." or, "Nobody told us about it." These are key words pointing to a lack of teamwork, communication, and coordination among the participating engineers.

Establishing a good team and good communications during systems engineering is essential to the process. The core of the engineering team should be the vendor's systems engineer, who consults on relevant items with the HVAC and facilities engineers. The engineering team can be supplemented by other specialists on an as-needed basis. These can be specialists for communications, environmental safety, production, or operations and maintenance.

The global commissioning/TQM approach makes systems engineering more complete because it requires implementing ideas from different aspects of engineering and operations. It also leads to higher quality due to engineering reviews at each significant stage of the systems engineering process.

The following is an example of a systems engineering team:

- Facilities Engineer
- HVAC Engineer
- Systems Engineer
- Other Specialists

The Systems Engineering Review Team

A proficient review team also should be established. Both teams are essential to the success of systems engineering and the review process.

The core of the review team is the commissioning agent. This individual assembles the review team and assigns individual tasks to the team members. Participation of operations and maintenance personnel is essential to the review team. Systems operators are the best resources for providing input to development and for reviewing of operators' interfaces (PMIs). They can also contribute to the definition of alarm limits, alarm messages, set point values, operating schedules, and other operational items. Operators are undoubtedly excellent resources of information for the software engineer.

Similarly, maintenance personnel bring real life experiences with the same or similar DDC systems to systems engineering. Their experience involves systems operation, component failures, locations of field gear, calibration, methods of troubleshooting, and other maintenance issues.

The commissioning agent must overcome initial objections to involvement of O&M personnel in the systems engineering process. The most common objection is that, "They should not be wasting their time with systems engineering reviews, they should be doing maintenance." Time spent on systems engineering reviews has a long-term impact on operations and maintenance. The owner benefits from operators' and maintenance technicians' involvement because it leads to their ready acceptance of the DDC system, and a full understanding of the controls and automation sequences for the designed building or systems. Having O&M input to software design (alarm limits, messages, set points, etc.) shortens the system's start-up time. Having their input to systems engineering (hardware selection, locations, etc.) shortens system's turn-over time. O&M personnel are well acquainted with site-specific issues, and their practical suggestions can improve DDC system definitions as they relate to systems operations and maintenance. Involvement of O&M personnel in the systems engineering process is beneficial for the project and for long-term systems operation.

The following is an example of a review team:

- Commissioning Agent
- Facilities Engineer
- HVAC Engineer

- Systems Operators
- Maintenance Technicians

Performance Requirements and Measures

Performance Requirements

The purpose of performance requirements for systems engineering is to aim at quality systems design and applications software development. The combination of engineering the most suitable hardware (field gear with their sizing and locations, DDC architecture, controllers and front end) along with application software assures control of the building environment within design parameters. The commissioning agent should base the performance requirements on job and site requirements and specifications, realizing that some requirements are best transferred among engineers during the review process. This fosters an open exchange of ideas and their acceptance by the systems engineers, which is essential for the process.

Performance Measures

Periodical quality assurance reviews during engineering and software development ensures a DDC system design which meets design requirements. A review of systems engineering tasks should be scheduled at completion of significant milestones. The following is a sample checklist of QA review stages during systems engineering.

Checklist of QA Review Stages

- ☐ Review of HVAC Engineering Submittal
- ☐ Review of Owner's Submittal
- ☐ Review of Instrument List
- ☐ Review of System Architecture
- ☐ Specification Review of Standalone Controllers
- ☐ Specification Review of Operator Interfaces
- ☐ Review of Communication Design and Testing
- ☐ Review of Application Engineering and Controls Design
- ☐ Review of Electrical Design
- ☐ Review of Application Software
- ☐ Witnessing of Software Testing
- ☐ Review of Person Machine Interfaces
- ☐ Documentation Review
- ☐ Factory Acceptance Testing

The commissioning agent should add to the checklist job-specific or site-specific items for a given project.

Systems engineering checklists can aid internal and external customers by providing a structured approach to the review process at important stages of systems engineering.

The extent of use of formal checklists in the review process depends on the job, the commissioning agent, and the quality of the design and review teams.

HVAC Design Submittal

Item Description	Reviewed/Approved (Y/N)	Not Submitted	Resubmit Date
1. HVAC specification			
2. HVAC systems schematics			
3. HVAC layout			
4. HVAC design parameters			
5. HVAC equipment sizing			
6. Non-HVAC equipment connected to DDC			
7. DDC job-related specification			
8. DDC point list			
9. Sequence of operation			
10. Requirements for DDC vs. local controllers for each HVAC unit			
11. Specification of HVAC equipment with third party dedicated controllers			
12. Non-DDC controlled equipment			
13. Location of DDC controllers			
14. Electrical specification			
15. Other job-specific items			

Some situations require adherence to formal checklists, in other cases the check-lists serve as a guideline. The owner's commissioning agent should not use the checklists as a form of bureaucratic process. The sample checklists are a compilation of important or significant systems engineering items.

Review of HVAC Engineering Submittal

Notes about HVAC Design Submittal checklist:

1. The HVAC specification (Division 15000) contains the Scope of Work for the DDC system.

2. HVAC systems schematics are one-line diagrams of air and hydronic distribution systems including fans, pumps, coils, and valves, their sizes, calculated flows, pressures, temperatures, and other relevant engineering information.

3. HVAC layout drawings identify equipment locations in the mechanical rooms, air duct and piping routing in the building, and locations of equipment controlled by the DDC system.

Note for items 2 and 3: Use of site acronyms, as well as final room numbering in the HVAC documentation, is important.

4. The description of design parameters for individual HVAC systems, zones, and rooms includes set points, and control and alarm limits for pressure, volume, temperature, humidity, and other controlled variables including required differentials. Parameters should be defined for normal, stand-by, and night set-back modes of operation. Include scheduling of major equipment (daily, weekly schedules, etc.), and start-up and shut-down sequences.

5. Show calculations of building heat gains/losses, air volumes, air changes, coil capacities, temperature rise/drop across coils, and other calculations

relevant to equipment sizing should be part of the submitted documentation. Include cut sheets of major equipment, or components relevant to DDC systems engineering, in the submittal.

6. Every building has other (non-HVAC) equipment that can be controlled and/or monitored by the DDC system, especially in facilities with central dispatching. This includes secondary monitoring of fire prevention equipment (dry pipes, valves, pumps, alarm status of monitored zones, etc.), security systems, access control, and operations-related equipment such as critical freezer temperatures, laboratory equipment, domestic hot water temperatures and pressures, sump pumps, and leak detection equipment.

7. In multi-building applications, the owner should provide a site DDC specification accompanied by a job specification written by the HVAC engineer. In single building applications, the HVAC engineer provides a DDC specification for the application. For more details on specifications refer to Chapter 2.

8. A DDC point list (I/O list) is an important guideline for the systems engineer. The HVAC engineer, facilities engineer, and the systems engineer finalize the point list (or instrument specification) during the systems engineering process. For example, the (HVAC) engineer may have specified DDC points for control of a mechanical room fan. However, the facilities engineer may prefer control of the fan by a local thermostat, and monitoring of the room temperature (high temperature alarm) by the DDC system. Another example is specification of certain chiller points for DDC systems monitoring. However, the selected DDC system may have an interface protocol to that particular chiller controller, making point mapping via communications protocol more cost effective. They have to decide between them which method to use.

   Flexibility of the engineers and a common desire to provide the best system for the application lead to optimum utilization of the DDC system and lowest cost for the owner.

   A good example of a point list is the Input/Output Summary Table published in the ASHRAE Guidelines 1-1989, reprinted in Chapter 2. Many consulting engineering firms and DDC vendors have developed their own formats of Input/Output summaries. If the facility has a standardized format, the consulting engineer, as well as the systems engineer, should follow it to aid in standardization of site documentation. I/O summaries should use naming conventions standardized for the facility.

9. The Sequence of Operation is a written description of the HVAC engineer's intention to operate the building environmental control systems. The description is considered preliminary until firmed up with the systems engineer and facilities engineer during the application software development phase of the project.

10. Depending on the controlled HVAC systems and DDC system's architecture, some equipment can be controlled by standalone controllers, unitary or equipment specific controllers with limited data base, or by full featured DDC controllers. Which combination is the most suitable for the application, and can be supported by the selected DDC system, depends on the situation evaluation by the HVAC, systems and facilities engineers. They should establish an optimum balance between systems cost and their reliability in case of DDC controller or communications failure. A full featured

building controller may cost in excess of $5,000, while an equipment specific controller may be in a price range of $500–$1,000. Utilization of less expensive controllers is appropriate for some equipment; other applications warrant utilization of "full" DDC controllers throughout the system.

11. Third party controllers associated with specific major mechanical equipment have become a fact of life. Such dedicated controllers may be controlling boilers, chillers, water treatment, ice-making, greenhouses, and other special equipment. The advantage of using dedicated controllers is in their association with vendor-specific mechanical systems. They were designed and tested on a specific system, therefore they are perfect for such equipment. Most major DDC manufacturers have developed interfaces (drivers) to the most common third party controllers. Before purchasing major equipment with a dedicated third party controller, ask the following questions:

    a. Does the DDC vendor have a (tested) communication driver to the third party controller?

    b. Was the interoperability between the third party controller and the DDC system tested with "exactly the same software, hardware, and communications scenarios" as are present on the job site?

    c. In the absence of an "industry standard" driver, who is responsible for its integration, performance, and upgrades, and what is the total cost for a unique "custom driver?"

    d. Is the use of third party controllers in compliance with site standardization?

    e. Will implementation of such controller(s) require extra training for operation and maintenance?

    f. How much integration do you really need (as opposed to "it would be nice to have")?

    g. Can hardwiring of essential alarms to the DDC system for reporting and monitoring functions be sufficient for systems operations?

12. Not all equipment in the building has to be DDC controlled. For example, a ventilation unit for a mechanical room operating 24 hours a day can be controlled by a local sensor or thermostat, and it does not require sophisticated (and more expensive) DDC controls. Critical points from locally controlled equipment (or areas) can be connected to the nearest DDC controller for monitoring and alarming of desired functions and conditions.

13. Space allocation for DDC controllers in mechanical rooms and other areas, along with suitable access and adequate lighting, is essential for DDC systems. The job documentation should provide the systems engineer with adequate access to, and lighting for, DDC controllers.

14. Electrical specification (Division 16000) related to AC power and DDC wiring, termination, labeling, and other coordination activities between the building electrical and controls contractors, should be part of the documentation submitted to the DDC vendor.

15. Amend this list with other job-specific items relevant for the job.

The commissioning agent reviews the quality of the documentation prior to turning it over to the DDC vendor. In the case of missing items from the documentation, or items requiring further clarification or approval, the commissioning agent and the project manager should evaluate the impact of these items on the quality of the process and overall job schedule, and set up deadlines for their re-submittal.

Review of Owner's Submittal

Notes about owner's submittal checklist:

1. Facilities, especially large and multi-building facilities, should develop site design standards. Published site standards are provided to design engineers and also to controls vendor's systems engineers. Site standards should include requirements for DDC architecture, the controls and automation master plan, site acronyms, systems interfaces, networking requirements, and other site-specific issues. In the absence of facilities standards, DDC systems engineering related issues should be determined between the facilities engineer and the HVAC engineer, and reviewed with the DDC systems engineer. In such cases, site-related issues should be included in the job documentation.

2. Site DDC specifications, such as Performance and Material Specifications, contain site requirements for DDC system design and implementation. In the absence of DDC specifications, related issues should be determined between the facilities engineer and the HVAC engineer, and reviewed with the DDC systems engineer. In such cases, site DDC specifications related issues should be included in the job DDC specification.

3. The scope of work for individual retrofit projects seldom includes systems interfaces. However, most facilities have BAS or other computerized facilities automation systems to which the new DDC system should interface to. Some facilities may require interoperability with other facilities au-

Owner's Submittal

Item	Description	Received/Approved Y/N	Not Submitted	Resubmit Date
1.	Site design standards			
2.	Site DDC specifications			
3.	Site requirements to interface to existing DDC systems and networks			
4.	Preferred communications methods within the building			
5.	Preferred communications methods external to the building			
6.	Requirements for operator interface to the DDC systems for control and monitoring			
7.	Requirements for systems access and security			
8.	Material standardization			
9.	Maintenance practices			
10.	Required energy management functions			
11.	Requirements for information transfer to other systems			
12.	Operating policies for central controls vs. local control and associated responsibilities			
13.	Connections to existing building systems/equipment utilities			
14.	Definition of alarm messages			
15.	Other site-specific issues and requirements			

tomation systems such as power plant control systems, utilities metering systems, third party control systems, and maintenance management systems. The systems may have their own proprietary networks, or may communicate on a common facilities network. Any such site requirements have to be specified for the DDC systems engineers.

For sites with ongoing construction activities, the owner's engineering organization should determine all essential interfaces to existing environmental control systems, networks, and other facilities systems. The facilities engineers should provide a communications master plan for the DDC systems engineer with required interfaces, communications protocols, available communications media, and other relevant information.

DDC systems pre-selected for the site should be compatible with site communications requirements. Many DDC systems on the market support an array of industry standard or third party communication protocols. Interfaces between different vendors' building automation systems can be implemented by using standard communications protocols such as Canadian Automated Building protocol (CAB), ASHRAE's BacNet, or by utilizing either industry standard or custom written protocols or drivers. The most suitable interface options, based on reliability, availability, speed of communications, and cost should be considered for the site.

Site interface and networking requirements should be reviewed by the facilities and systems engineers.

4 and 5. Every DDC vendor offers several communications options with their systems. The owner's facilities engineers define the most suitable communications methods for the site and buildings.

These should consist of the following definitions:

a. Required communications speed (9600 baud, 20Mbps, etc.)
b. Communications networks (Ethernet, Arcnet, RS 485, etc.); their topography (bus, star, etc.); use of communications hubs, repeaters, bridges, etc.; protocols (proprietary, CAB, BacNet, etc.); and drivers (Allen Bradley Data Highway, Modbus, etc.)
c. Communications wiring (number of pairs of twisted shielded wires, dedicated telephone lines, existing telephone lines for dial-up modes, coaxial cables, fiber optic cables, etc.).

6. The operator interface to DDC systems for control, monitoring, and maintenance should be defined by the owner. Many multi-building facilities have central dispatching or control centers. Such centers monitor and operate building environmental control systems 24 hours-a-day year round. More and more occupants of multi-building facilities require local monitoring of their buildings by their facilities or business offices.

To what extent the operator interfaces are centralized or distributed is not an engineering issue because most DDC systems can have multiple OWS on the network. But facilities have to decide on the number of OWS based on operational requirements. Operations considerations include: location of OWS for central control, distribution of OWS for customers (local access), assignment of work stations for maintenance personnel, access levels for all OWS having access to the DDC network, assignment of responsibilities to respond to system and building alarms, and assignment of overall responsibilities for systems operation and maintenance. These responsibilities should be defined for every OWS on the network.

In assigning responsibilities to access a real-time system, and to operate live equipment, consider the following:

a. Systems integrity

b. Assignment of responsibilities for environmental control systems operation, response to alarms, response to complaints from the building's occupants, safety of maintenance personnel working on associated mechanical systems, and responsibilities for DDC system software maintenance

c. Format of data presentation on individual OWS and frequency of updates, what information should be presented on individual OWS (all, related to HVAC systems, or only related to environment, such as temperature, humidity, etc.); alarms presented at individual OWS, their format, and policies related to their acknowledgement and operator's response.

7. System access and security levels have to be determined for each DDC controller and OWS to preserve the integrity of the system. There is an inherent paradox associated with modern systems architecture: the friendlier the systems get, and the more systems are linked on one network, the more vulnerable they become to intentional or unintentional access. Networking without systems integrity and security considerations may become detrimental to systems integrity, and could jeopardize the safety of connected mechanical systems and the people working on them. Unlike data processing systems and networks, real time systems and networks control live equipment and utilities (such as steam, hot water, chilled water, or electricity). Systems integrity and building systems safety issues must be of primary importance for DDC systems engineering.

Systems access and security issues are often outside of the design considerations (especially when the intention is to select the lowest cost DDC system). However, in the real world, there are computer viruses, hackers, or people who (intentionally or unintentionally) can start (or shut off) a large chiller from a remote terminal somewhere on the network. The implications of such "unintentional" actions are catastrophic enough to ruin anybody's day and teach a lesson or two about systems security.

8. Standardization in preselected systems for a site is important for larger facilities. There are more DDC systems on the market than can ever be implemented on one site, even in large facilities. Those brave enough to attempt implementing, for example, more than half a dozen different systems for one facility, would be creating a controls and automation systems "tower of Babel." The complexity of systems would defeat the main reason for automation, which is to contain operating and energy costs. Despite open protocols and networking, facilities have to standardize with a definite number of systems for their sites. Industry standard protocols and availability of interfaces make the issue seem less important. However, implementation of several different systems and their upgrades over several years can create an unmanageable situation for the facility. The problem arises in the need for expensive interfaces; the purchase and stocking of many different hardware components; the lack of comprehension of different systems software by engineers, operators, and maintenance personnel; and in increased systems maintenance costs. Standardization, on the most appropriate DDC systems and controls hardware, should be a goal for every facility.

9. Every facility has specific standard maintenance practices. Some maintenance crews are unionized, some are not. Some facilities have maintenance crews capable of diagnosis, troubleshooting, and repair of the most intricate systems related problems. Other organizations rely solely on the DDC or other contractor's services for maintenance and repair. For large multi-building, multi-(DDC) systems facilities, developing in-house DDC system expertise not only makes sense, but saves money in the long run for systems implementation, operation, and maintenance. Site-specific maintenance practices should be discussed up-front with the controls vendors or contractors. In return, maintenance requirements associated with proposed DDC systems should be discussed with the owner's O&M management, especially if they would result in increased manpower or restructuring of the existing maintenance department.

10. In the not-so-distant past, DDC systems were called energy management systems. Their greatest contribution to the evolution of building systems was in energy management features. Many of these typical features used in the late 1970s and early 1980s are no longer used to the same extent. Nevertheless, they represent an understanding common to owners, the HVAC design community, and DDC systems programmers.

Before submitting a long list of energy management features—copied from a 1980s energy management system's specification—be clear whether or not all features are needed for the building's operation. It does not mean abandoning all energy management features of the '80s, but being selective as to their applications and impact on occupancy comfort, productivity, and the optimization of operating costs for buildings. Energy conservation was and will be a big selling point for building DDC systems. Take full advantage of the DDC system's energy management capabilities, and build on the wealth of knowledge accumulated over the past decades.

11. With systems interoperability comes responsibility for information management. The owner should clearly specify the information required at each DDC controller or OWS, sources and recipients of information, time and frequency of information transfer, and whether the information is needed for human processing or for controls.

Utilizing communication protocols and/or drivers, all points (hardware and software points) from one system can be mapped over to another system. For some applications such practice is not necessary. To optimize information flow and related speed, the owner has to decide which information (points) should be mapped over to the other system, how the information is going to be used and by whom, and at what frequency (scan rate) the transaction should occur.

The owner's facilities engineer and the DDC systems engineer have to determine the impact of information transfer on the DDC system's performance. They have to examine whether or not the volume of transferred information will create a bottle-neck on the communications system, slow down the scan rate, slow down the screen upgrade, or show other negative effects that a massive data transfer could have on the installed DDC system's performance. There should be consideration for human recipients of transmitted information as well—operators, maintenance personnel, and other personnel. Information that is not processed (by systems or humans), or responded to (e.g., field alarms reported by DDC systems), is meaningless for the user as well as systems operations. A multitude of such "unnecessary" information can cause degradation of systems performance, and at the least, will result in the extensive use of printer paper. Probably, if no one looks at the presented information, no one needs it.

12. Expectations from building occupants regarding systems operation, local monitoring, interfaces, and equipment scheduling, should be brought to the DDC vendor's attention during this stage of the project. Data presentation, access to data, level of control and manipulation of set variables by local operators, and their responsibilities, should be determined by the facilities engineers.

13. Equipment not specified by the HVAC engineer that needs to be controlled or monitored by the DDC system. Because such equipment is an addition to the project scope and DDC point list, it should be specified early enough to include it into the scope and systems engineering.

14. Some alarms reported by the DDC systems have associated alarm messages. An alarm message can be related to systems performance; or a combination of events can trigger a message advising operators to initiate an action, or to determine a trade's response to the reported HVAC problem. Such site-specific requirements should be brought to the DDC systems engineer's attention at this stage of the project.

This checklist for the Owner's Submittal should be modified by site-specific issues, to assure compliance with site requirements.

Review of Field Instrument List

The Field Instrument List should be completed, reviewed, and approved as early as possible because some instruments have long delivery schedules. With the site requirements for standardization in mind, the systems engineer should design instruments in accordance with site specifications, design criteria, or material specifications. In the absence of such documents, the systems engineer should review the proposed field instrumentation with the owner's facilities engineer and HVAC engineer. The following format can be used for building DDC controls as well as for industrial controls applications (such as power plants):

Field Instrument List

Acronym	Used for	Instrument Range	Operating Range	Voltage Amps Pressure	Make	Model	Remarks
Example: AH1CVL	Cooling coil valve control	4-20 mA	0-100%	24VAC	XYZ	123	Electrical actuation

Review of DDC System Architecture

DDC system architecture should match the physical layout of the buildings, HVAC systems, and controlled points. The location of field controllers should match the point concentration in the building and mechanical systems. Field controllers should be located at the highest concentration of field points to save on wiring costs. The selection of field controllers should comply with the owner's requirements for remote operation, standalone capabilities, and other site requirements. Utilization of inexpensive unitary or equipment-specific controllers connected to building controllers may be the most economical architecture for some installations. Other installations may require use of more expensive full-featured DDC controllers for every mechanical room. The DDC systems engineer, in cooperation with the owner's facilities engineer and the HVAC engineer, should engineer the most suitable (and economical) architecture for the application.

The following is an example of a checklist for DDC systems architecture:

DDC System Architecture

Item Description	Approved/ Not Approved	Not Submitted	Resubmit Date
A. Building controllers			
1. Utilization			
2. Location			
3. Connected field points			
4. Grounding			
5. Surge protection			
6. Application programs			
7. Weekly scheduling			
8. Energy management programs			
9. Alarm handling			
10. Communications software			
11. Communications interface			
12. Transmission speed			
B. Unitary/Application-Specific Controllers			
1. Utilization			
2. Location			
3. Connected field points			
4. Grounding			
5. Surge protection			
6. Application programs			
7. Weekly scheduling			
8. Energy management programs			
9. Alarm handling			
10. Communications software			
11. Communications interface			
12. Transmission speed			
C. Third Party Controllers			
1. Utilization			

DDC System Architecture—continued

Item Description	Approved/ Not Approved	Not Submitted	Resubmit Date
C. Third Party Controllers			
2. Location			
3. Connected field equipment			
4. Grounding			
5. Surge protection			
6. Application programs			
7. Weekly scheduling			
8. Energy management programs			
9. Alarm handling			
10 Communications software			
11 Communications interface			
12. Transmission speed			
D. Interoperability			
1. Interface to other building controllers			
2. Interface to other systems on site			
3. Communications networks			
4. Information access			
5. Point mapping/display			
E. Operator Work Stations			
1. Central OWS			
2. Remote OWS			
3. PMI screens/reports			
4. History data files			
F. System Security/Access			
1. Access at controller level			
2. Access at OWS level			
3. Password protection			
4. Network protection			
G. Site-Specific Items			

Review of DDC Controller Specification

DDC controllers are systems- or vendor-specific. The DDC systems engineer defines each controller based on required features. A specification of "generic" controllers by an (HVAC) engineer could add unnecessary costs to the project. The DDC controller specification should be reviewed by the HVAC and facilities engineers to assure compliance with the systems architecture, system reliability, controllers' standalone capabilities, and operation and maintenance requirements at optimum cost. The controllers should be designed with expansion in mind. Some controllers are designed with a predetermined number of digital and analog input/output modules; other controllers are modular. Future expansions can be achieved either by adding (spare) input/output modules, or by adding controllers on the same network(s). The method to choose depends on the individual DDC system and its interface capabilities.

The following is a sample checklist for DDC controller specifications:

DDC Controller Specification

Controller ID: **Location:**

Item	Description	Approved/ Not Approved (Y/N)	Not Submitted	Resubmit Date
1.	Enclosure			
2.	Power source			
3.	Battery backup			
4.	Wire terminations			
5.	I/O modules			
6.	Pneumatic outputs			
7.	Spare I/Os			
8.	Regulated power source			
9.	CPU/memory			
10.	Application programs			
11.	Energy management programs			
12.	Weekly scheduling			
13.	Alarm handling			
14.	Set point/alarm limit modifications			
15.	Software upload/download			
16.	Communication software			
17.	Communication interfaces			
18.	Data transmission rate			
19.	Operator interface			
20.	Diagnostics/reset			
21.	Surge/lightning protection			
22.	Other controller-specific items			

Review of Operator Workstation (OWS) Specifications

Depending on the owner's requirements and the system architecture, the project may require one operating workstation (OWS), or multiple OWS. Some projects are designed for existing facilities with OWS already in place. Some multi-building facilities, in addition to the existing central OWS, may require additional OWS. Any time the scope calls for new or additional OWS, the DDC systems engineer along with the owner's facilities engineer should consult the future operator(s) on their needs, operating practices, and preferences.

With enhancements in DDC distributing processing, the earlier front-end computers (PCs or mini-computers) lost their significance as "central processors" due to the distribution of application programs in the remote standalone controllers. Most OWS reside on a DDC network as "nodes" and are used as Person Machine Interfaces (PMIs). As such, they fulfill an important role in providing an interface to data residing in individual controller databases throughout the network. With fewer computing requirements and increasing power and speed, they provide "friendly" Person Machine Interfaces for the operators. In most systems, the operator interface is via colorgraphics screens of controlled HVAC or building systems, with real time data scanned and updated on the screen (CRT). Another feature of the PC-based OWS is their report generating feature, using off-the-shelf statistical programs, spreadsheets, and graphics software for management reports. Besides report generation, the operator interface PCs can be used for storing and manipulating history files, and providing a statistical analysis of building and HVAC systems performance.

DDC vendors have spent quite a bit of time and money researching the best combination of PCs and printers to provide the best PMI presentation of their DDC system. If feasible, the owner should implement front-end hardware and software tested and recommended by the DDC vendor.

The following is a sample checklist for OWS specifications:

OWS Specifications

OWS ID: Location:

Item	Description	Approved/ Not Approved (Y/N)	Not Submitted	Resubmit Date
A.	**PC Hardware**			
1.	Microprocessor			
2.	Base RAM			
3.	Cache			
4.	Expanded RAM			
5.	Hard Disk Drive			
6.	Access Time			
7.	Floppy Disk Drive(s)			
8.	Photo CD ROM			
9.	Local IDE Bus			
10.	ISA Expanded Bus			
11.	Local SVGA Bus			
12.	Graphic Accelerator			
13.	Parallel Port(s)			
14.	Serial Port(s)			
15.	Communication expansion card(s)			
16.	CRT (screen)			
17.	Keyboard			
18.	Mouse			
19.	Mass memory drive			
20.	Other			
B.	**Software**			
1.	Operating system			
2.	Application software			
3.	Graphic software			
4.	Statistical software			
5.	Communication software			
6.	Protocols			
7.	Drivers			
8.	Other third party software			
9.	DDE			
C.	**Printer(s)**			
1.	Paper width			
2.	Speed (cps)			
3.	Color			
D.	**Communication Interfaces**			
1.	Interface cards			
2.	Communication modems			

Review of Communications Design, Information Transfer, and Protocol Testing

Networking, common or industry standard protocols, and drivers added a new dimension to the already complex world of controls and automation. Just about everything related to these is complex and new to automation engineers. In addition to the application engineering of DDC systems, there are two added engineering disciplines the systems engineers have to master:

1. Communications and networking

2. Information transfer and management

Anyone who attempts to downplay the complexity and significance of these disciplines to DDC system applications can make costly mistakes. The costs of such an assumption can come, like all bad news, in many forms:

1. The facility can be strapped with several incompatible systems, adding costs and burdens to the operating and maintenance budget, and to personnel.

2. To make incompatible systems interoperable can be a giant undertaking, and a costly engineering task with limited success.

3. The transfer of an unlimited volume of information from remote DDC controllers over the network can slow down the connected systems.

4. Indiscriminatory reporting of every alarm and change of state on every OWS can overwhelm the operators.

5. Incomplete and inconsistent management reporting and information transfer of history data over the network can jeopardize data validity and integrity.

6. A lack of network and systems security can jeopardize the integrity of the entire network and connected mechanical systems.

Real uncharted territories lie outside of the DDC vendor's proprietary protocols. Any facilities engineer venturing outside of these boundaries should take a serious look at the needs of the facilities, future systems development, and the structure of the facilities operations and maintenance. Proceed cautiously and implement field-proven (as opposed to factory tested) communications and interface solutions. Field-proven applications in operation for an extended time (months or years) should be researched as to site implementation experience, reliability, ease of installation, and start-up. The best way to learn about operating experience with installed systems is from the operating engineers at the site. If, for any reason, it becomes inevitable to install interfaces and protocols which are not operational in a similar field environment, set up site tests on the owner's premises to test the proposed solutions. Always allow extra time and money for testing! Interfaces and communication protocols seldom work the first time around.

The following is a sample checklist for communication design, information transfer, and site testing of communications protocols.

Communications Design, Information Transfer, Site Testing

Item	Description	Approved/ Not Approved	Not Submitted	Resubmit Date
A.	**Third Party Protocol**			
1.	Point definitions			
2.	Points to be mapped over			
3.	Scan rates			
4.	Control logic access			
5.	Network access control			
6.	Protocol services			
7.	Broadcast services			
8.	Review of literature			
9.	Conformance testing			
B.	**Third party drivers**			
1.	Point definitions			
2.	Points to be mapped over			
3.	Scan rates			
4.	Control logic			
5.	Review of literature			
6.	Conformance testing			
C.	**Field testing**			
1.	Communications testing			
2.	Protocol functional testing			
3.	Communication speed			
4.	Error rate			

Review of Application Engineering and Controls Design

The design of temperature controls for the building industry evolved from the design of local single loop pneumatic controllers in centralized energy management systems to distributed DDC systems. While the early forms of applications engineering and controls design were strictly hardware oriented (controls schematics and ladder diagrams), DDC controls design consists of two parts:

1. Design and specification of hardware—DDC systems hardware and field gear associated with the application

2. Design of application software—software definition of points, and definition of operating logic.

DDC vendors and consulting engineers use different forms of graphical presentation of controls design. The controls design is always based on the mechanical layout system (e.g. AHU) and associated hardware points.

For multiple building facilities using multiple DDC vendors, the controls design has to follow the owner's requirements for uniform design and documentation. It may be easier said than done because each vendor, consultant, or contractor is using their own standard documentation and library of symbols developed over the years. Therefore, it is an enormous task to convince the vendors and engineering organizations of the need to standardize their documentation to site requirements.

However, vendors and engineering organizations should adopt the owner's requirements. They should understand that each of the participating organizations is involved in relatively small sections of the design, as for example, in the renovation of an AHU, one or several floors of the building, or construction of a new building for a multi-building facility.

Years of lack of standardization on engineering documentation and submittals catch up with many facilities. One look at the documentation archives of a multi-building facility speaks for itself. Many multi-building facilities spend large amounts of money on the consolidation of existing documentation. With the ever-growing numbers of building systems being installed and facilities being automated, multi-building facilities cannot continue accepting engineering submittals and as-built documentation that do not meet site standardization criteria. Owners are paying customers, and the involved organizations should provide the owners with documentation meeting site standardization requirements.

Standardization of DDC documentation is critical because the DDC documents are used daily by operations and maintenance personnel. They are manuals that are used daily by:

a. Maintenance departments for hardware and software calibration, troubleshooting, and repair and analysis of building or HVAC problems
b. DDC operators for systems diagnostics, software modifications, building problem analysis, and as reference material
c. Facilities engineers responsible for long-term systems development, upgrades, networking, and interoperability of all computerized systems used at the facilities department.

One of the essential items to be standardized and utilized by all vendors providing DDC systems for the facility is site naming conventions. The same naming convention should be used in all systems (point names, acronyms, etc.) to create unique point names throughout the facility. With the availability of open communications protocols, and in an effort to optimize the facilities resources, one common OWS can interface to all DDC systems on the network. With that, uniform data presentation, graphics screens and uniform point naming conventions (regardless of which DDC systems they come from) become a necessity.

The commissioning agent should also assure compliance with facilities requirements on delivering the documentation on certain software, such as Auto-Cad, WordPerfect, Lotus, and others commonly used at the site. This is an important requirement because many facilities have transferred their existing site documentation to electronic media that can be accessed by engineering and O&M personnel over the facilities network.

The following checklist can be used for application engineering and controls design:

Application Engineering and Controls Design

Item	Description	Approved/ Not Approved (Y/N)	Not Submitted	Resubmit Date
1.	HVAC system control drawings			
2.	DDC controllers design			
3.	DDC points list			
4.	Naming convention			
5.	Instrument lists			
6.	Instrument ranges			
7.	Normal and failsafe positions of dampers and valves			
8.	Wiring and wire terminations			
9.	Wire labeling			
10.	DDC controller I/O modules			
11.	Surge protection			
12.	Communication ports			
13.	Grounding of controllers			
14.	AC supply			
15.	Location of controllers			
16.	Pneumatic interfaces			
17.	Hard wired interlocks			
18.	Fire safety interfaces			
19.	Sequence of operations			
20.	Communications configurations			
21.	Instruction manuals			
22.	Conformance of engineering documentation to site standards			
23.	Conformance of software documentation to site standards			
24.	Conformance of O&M instructions to site standards			

Review of Electrical Design

The project manager determines the most economical ways to coordinate electrical installations for the project. For some jobs, the building electrical contractor is also responsible for DDC wiring. Other jobs may have a second electrical contractor (subcontracted by the controls vendor) providing installation of controls wiring. The project must have clear specification of responsibilities for electrical wiring between the controls and building electrical contractors.

Generally, building electrical contractors are more proficient in AC power wiring than in point-to-point controls wiring. The priorities of site electrical contractors are guided by the building installation schedule. This schedule may compete for the same time frame as installation of controls wiring.

The following tasks, related to electrical installations and wiring, have to be coordinated for DDC systems:

1. AC power sources for controls field cabinets

2. AC power sources for field actuators and other devices

3. Low voltage and AC power wiring between DDC controllers and installed field points

4. Communications wiring

5. Systems grounding

6. Wire labeling

7. Wire termination

8. Polarity testing

Most control vendors and contractors are capable of complete electrical installations of their systems. The advantage of this approach is that it becomes a "turn key" job for the controls contractor. All responsibilities for job scheduling and coordination of construction activities are in the hands of the controls contractor. If the controls wiring is installed by the electrical contractor who is subcontracting to the controls contractor, the controls contractor has to coordinate all wiring activities with the building electrical contractor (connections to the building electrical panel for AC power, wire routing, etc.) and construction manager.

Another method often chosen by the project managers is for the building electrical contractor to install controls wiring as part of the job (assuming the electrical contractor has experience with controls installations). Coordination of activities between the job electrical contractor and the controls vendor is more demanding than it would be in a "turn key" situation.

Besides coordinating individual tasks between the two contractors (i.e., electrical contractor will install electrical wiring, and will provide wire and cable labeling and testing, and the controls contractor will provide wire termination), the project manager has to coordinate job schedules overall, which is more demanding. The project manager may encounter unforeseen delays and expenses, for example, discovering incorrect labeling during field testing by the controls technicians, or scheduling controls installations prior to the completion of wiring.

The following checklist can be used for review of electrical documentation associated with the DDC system.

Review of Electrical Design

Installation by DDC Contractor:

Installation by Building Electrical Contractor:

Item	Description	Approved/ Not Approved (Y/N)	Not Submitted	Resubmit Date
A.	**AC power supply for each controls field cabinet**			
1.	Power source			
2.	Amperage of the AC breaker			
3.	Protective guard on the breaker			
4.	Labeling			
5.	Grounding			
B.	**AC power supply for each field actuator and other devices**			
1.	Power source			
2.	Amperage of breaker			
3.	Protective guard			
4.	Labeling			
5.	Grounding			

Review of Electrical Design—continued

Item	Description	Approved/ Not Approved (Y/N)	Not Submitted	Resubmit Date
C.	**Wiring between controllers and field points**			
1.	Wire size			
2.	Cable type			
3.	Labeling			
4.	Installation method			
5.	Shielding			
6.	Grounding			
D.	**Communications wiring between controllers in the building and outside the building**			
1.	Wire size			
2.	Cable type			
3.	Labeling			
4.	Installation method			
5.	Shielding			
6.	Termination			
7.	Balancing/end of line terminations			
8.	Grounding			
E.	**Design of labeling and testing**			
1.	Cable labeling			
2.	Color coding			
3.	Wire testing			
4.	Wire labeling			
F.	**Wire termination**			
1.	Types of terminals			
2.	Spare capacity of terminal blocks			
G.	**Testing requirements**			
H.	**Compliance with codes and standards**			

Review of Application Software

Development and review of the application software is one of the most significant systems engineering tasks for automation of environmental control systems and building functions. Application Software provides:

- Point definition
- Operating logic definition
- Alarm reporting definition
- Set point definition
- Trend, history, etc. definition
- Reporting

- Problem analysis routines
- Communications routines, etc.

Unfortunately, application software development is left entirely to the systems programmer in most situations. Engineers and project managers are preoccupied with other tasks. There is, usually, limited time to provide input, discuss ideas, or review the application program.

The importance of operating logic is undervalued and limited to the sequence of operation by most design teams. At present, there are more checks for hardware specifications than there are for the operating logic. However, experience shows that it is easier (and less costly) to purchase an additional piece of hardware than it is to correct inadequate operating logic that is causing poor systems performance. This "traditional" approach may have its roots in the design of single loop pneumatic controllers. There, the degree of automation was limited by the function of single loop controllers, with one control output, one input, and a set point.

DDC systems are different with their "open" software design, which allows for the design of varying degrees of automation functions, energy management features, monitoring of field points, alarm reporting, and generation of reports and advisories for building operations. The DDC systems programmer can program the application software for various degrees of automation from defining individual control loops through providing set point adjustments based on field input, operator input, mathematical calculations, software look-up tables, and cascading of control loops, and adding other advanced programming for efficient systems control. How much of the available DDC capabilities are utilized for individual jobs depends on the engineers and the systems programmer defining the operating logic in the application software. It is true that a system is as good as the operators operating it, but it is also true that the application program and automation of building systems is as good as the engineers developing it. A well-defined operating logic along with a well-defined HVAC system will determine the environmental system's performance under all operating conditions. A well-defined application software will also provide the level of desired automation for the building.

The process of defining control routines and operating logic is very creative. It should be done jointly by the HVAC engineer, facilities engineer, and the DDC systems engineers. Each of the engineers can contribute to the development of the application software with input based on specific needs and personal experience. Together, they can create an operating logic that will assure efficient systems operation and will be easy to operate and maintain.

The roles of the team members are determined by their expertise. For example:

- The HVAC engineer describes the sequence of operation of the HVAC systems based on the HVAC systems' design, required design parameters, calculated systems performance, and building characteristics.

- The facilities engineer provides site-specific requirements related to alarm reporting, report generation, requirements for on-line building and HVAC systems and their problem analysis, required operator advisories, requirements for system access and security, operator interfaces, and other site-specific requirements. The facilities engineer's input is based on past practices (what had worked in similar situations in the past), on facilities requirements for systems operations, operating policies and practices, structure of the maintenance department, response to alarms, customer complaints, and systems networking requirements. Energy management and energy reporting requirements should also be defined by the facilities engineer. In the absence of a systems integrator, the facilities engineer should define requirements for future automation systems development and DDC interfaces.

- The application engineer's task is to convert the requirements of the other engineers to an operating logic, and consequently, to a DDC program. Most DDC systems engineers and programmers are not only knowledgeable about DDC systems engineering and programming, but also have a wealth of knowledge about automation routines related to HVAC and building automation. They base their control routines on experiences in previous jobs, and on the library of control routines developed by other engineers within their organization. Systems programmers also use specific programming tools which aid them in defining individual control loops, control routines, and to test their programs prior to loading them into the DDC controllers.

Definition of control loops, the most prominent systems engineering task of past generations of controllers, is becoming more and more automated as software tools become available. More and more, the attention and time of systems engineers is devoted to programming advanced automation routines, on-line problem analysis, building operator interfaces, defining communications interfaces, and defining networking and systems reporting features.

The process of application programming is one of the most creative of engineering activities. It covers the whole spectrum, transferring intentions, concepts, and visions into application programs that provide automated operation of the controlled building systems, manage energy, and aid in efficient operation and maintenance of building systems.

Effective applications engineering translates into optimization of operating costs (energy and O&M costs) for the owner. While the energy savings features of BAS are well understood and used for project justification, savings on O&M costs due to proper application engineering is much less understood, acknowledged, and considered for project justification. This is due to a lack of methods for evaluating performance and savings on operations and maintenance costs over the life span of the system, and because DDC projects are often justified on a first cost basis. However, selecting the lowest cost DDC system and inadequate systems engineering will more likely result in increased operating and maintenance costs for the facilities.

To illustrate the effect of the quality of application software, consider a simple control of a heating zone (room):

1. A simple zone control loop:

   - Room sensor measures the room temperature
   - Set point is set manually (from the keyboard)
   - Valve is controlled by PI control loop
   - Deviation from the set point is reported as low or high alarms.

2. DDC automation of the same zone control loop:
   a. Valve control:

- Room sensor measures the room temperature
- Set point is determined for occupied/standby/unoccupied modes of operation (or manually from the keyboard)
- Based on differential between the room temperature and set point, outside air temperature, and occupancy time, the optimum warm-up time is calculated
- To prevent "hammering" in the cold steampipes (due to introduction of full steam pressure and temperature to the cold pipes containing residual condensate), gradual non-linear opening of steam valve over time is calculated. The valve control sequence and the time required for full valve opening is modified based on the outside air temperature. The calculated time delay for full valve opening influences the optimum calculated time for warm-up in the morning (optimum start), and introduces a "time leg" for the PI control.
- Calculations and logic provide input for the PI controller

b. Reporting to the OWS:

- The room sensor reports the analog value (room temperature), but also unreliable status, open wire, drift, and low and high alarms.
- Alarm limits and alarm messages are modified according to occupied/stand-by/unoccupied modes of operation
- If there is a low steam pressure alarm (reported by a steam pressure sensor), the information is transferred to the heating plant and an alarm message is generated for maintenance personnel
- If the condensate temperature of the heating zone is greater than the defined temperature (i.e., 190°F) for over 30 minutes, an alarm related to the condensate system failure is generated for maintenance personnel
- If the condensate pump of a related condensate receiver is "on" for an extended time (e.g., 5 minutes) an alarm related to pump or condensate receiver failure is generated for maintenance personnel
- If the DDC system calls for opening of the steam valve, and the valve feedback shows no change (and the room temperature is below its set point), the DDC generates a valve failure alarm for maintenance personnel
- If the control loop is constantly overshooting, causing deviations of room temperature, the DDC system provides self-tuning of the PI loop, or reports the problem to the OWS
- All alarms are logged for statistical evaluation
- All alarms are acknowledged and transferred to maintenance management systems for work orders to be generated.

This example of zone value control demonstrates the degree of automation any DDC system can provide with the same hardware configuration. How much automation gets implemented depends on the engineering team and application software design.

The experience and ingenuity of engineers participating in the application software development can make a difference between a "simple" room control and an "advanced" automated process. While the first controls strategy provides adequate room control under most normal operating conditions, the advanced algorithms will result in controlling the room temperature under all operating conditions. On-line problem analysis and notification of O&M personnel will result in faster response to problems and reduced maintenance costs. Most importantly, it may save some important research or production in the controlled room.

Most DDC systems on the market have the analytical capabilities as described earlier. Whether or not they are utilized depends on the engineers defining the operating logic.

The following case study illustrates this situation: An AHU with several VAV boxes served the laboratory area of a building. The DDC system had been installed and turned over to the owner. During some cold wintry days, an operator expecting a forecasted extremely cold night temperature manually increased the preheat coil set temperature. In a few days the cold front had passed, the outside air temperature increased, and the laboratories were overheating. After many complaints from laboratory researchers, the operator called the DDC contractor's field technician. After about an hour of troubleshooting from the front end OWS, the technician found the higher set point of the preheat coil. He diagnosed the problem as an operator error, and claimed that the DDC system was working as designed; the overheating was caused by the manual increase of the original set point.

A follow up review of systems reports (alarm logs) revealed that none of the rooms reported high alarms even though the readings were in the 80s. There were no alarm limits defined for the room temperatures. The fan discharge temperature alarm was defined at high temperature, probably to eliminate nuisance alarms during start-up. There were several operators at each shift having unlimited access to the software. There was no record of operators logging on or of modifying set points—operator access and levels were not defined. There were only basic control loops, poor alarm reporting, and no trends defined in the system.

Any of the DDC parameters properly defined in the application program could have prevented the laboratory from overheating, or at least could have alerted the operators of building conditions. The DDC system had all the necessary features for automation, and the related field hardware had been installed. The situation resulted in financial losses to the owner due to the loss of laboratory experiments, and increased energy and maintenance costs.

Conversations held following the incident turned into one of the most common arguments between end users and DDC contractors: the end user blames the problem on the DDC system, and the DDC contractor insists that the DDC system is functioning as designed.

In remembering all advertized features and capabilities of the purchased DDC systems, the owner's fundamental question is, "Did we get the DDC system performance we expected and paid for?" In the case study, the HVAC engineer provided the DDC vendor with the control sequence. The DDC contractor defined the control loop of the heating coil with a minimal approach. What the contractor failed to do was to automate the process! The claim of the contractor that "the DDC system was working as designed," shows a great deal of ignorance, disrespect for the value of the owner's purchase, and a lack of system definition and teamwork. The owner purchased a DDC "automation" system, not a "simple" single loop controller, and paid a premium for the purchase!

The DDC vendor's system engineer did what he used to do with single loop controllers: defined an input, an output, and the set point of the control loop. All available advanced features of the DDC system were ignored. Alarm reporting was either not defined, or the alarm limits were set to the levels at which they became ineffective. Access levels for the operators and trending and report generation features were omitted. The overheating problem was reported in the old-fashioned way, over the phone by laboratory researchers. Then maintenance failed to respond to the problem on time because they focused on troubleshooting of the DDC problems, not the actual overheating problem. The DDC system became not only ineffective, but an extra burden for the maintenance personnel.

This example shows the importance of automation, teamwork, and level of competence for writing application software. However, knowing how to write a program does not substitute for the knowledge of controls and automation processes necessary to automate building environmental control systems. DDC systems specification, teamwork, and quality assurance throughout the entire process, and an extended performance warranty are what provide the owner with necessary checks and balances.

Software programs, in general, are difficult to review by persons not familiar with the particular program. In an effort to make programming more "user friendly," many DDC systems on the market have adopted forms of graphic programming. After learning the few basic rules of the graphic programming language being used, and getting familiar with the programmer's style, non-programmers are better able to review application programs.

Figure 6-1 is an example of a graphics programming language.

The application software should be reviewed for:

a. Each control loop

b. Each unit (fan, VAV, etc.)

c. Each system (HVAC, VAV, fume hoods, chill water, hot water systems, serving the same zone or area)

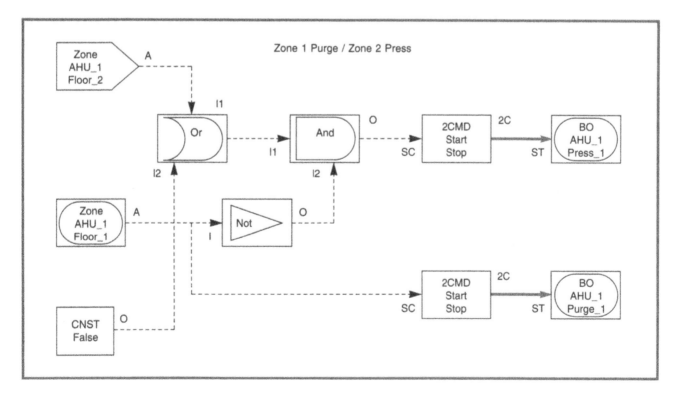

Figure 6-1. Example of a graphic programming language (GPL)

d. Each building
e. The operators' point of view
f. The maintenance point of view
g. The systems management point of view
h. The networking point of view.

Application Software

Item	Description	Approved/ Not Approved (Y/N)	Not Submitted	Resubmit Date
A.	**Each Control Loop**			
1.	P, I, or D			
2.	Loop constants			
3.	Input/output parameters			
4.	Space/coil parameters			
5.	Set point origin and definition			
6.	Alarm and message definition			
7.	Loop stability			
8.	Analog calibration			
9.	Points to monitor			
10.	Points to interlock			
11.	Diagnostics			
B.	**Control Algorithms for Each Unit**			
1.	Operating logic			
2.	Cascading of loops			
3.	Other input/output			
4.	Related calculations, tables, etc.			
5.	Operator advisories			
6.	Diagnostics			
7.	Alarm limits and messages			
8.	Winter/summer operation			
9.	Occupied/unoccupied modes of operation			
10.	Equipment scheduling			
11.	Energy management features			
12.	Back-up or failsafe operation			
13.	Start-up operation			
14.	Other specific conditions			
15.	Communications interfaces			
C.	**Each System Serving a Particular Zone, System or Building**			
1.	Definition of zones			
2.	Definition of systems in zone			
3.	Air balancing requirements/set values			
4.	Heat balancing requirements/set values			
5.	Zone parameters			
6.	System interlocks (hardwired and software)			
7.	Calculation of operating parameters			

Application Software—continued

Item	Description	Approved/ Not Approved (Y/N)	Not Submitted	Resubmit Date
C.	**Each System Serving a Particular Zone, System or Building**			
8.	Operator advisories			
9.	Diagnostics			
10.	Alarm limits and messages			
11.	Winter/summer operation			
12.	Occupied/unoccupied modes of operation			
13.	Equipment scheduling			
14.	Energy management programs			
15.	Back-up and failsafe operation			
16.	Start-up operation			
17.	Other specific conditions			
18.	Communications interfaces			
19.	Point mapping to/from controllers			
D.	**Each Building**			
1.	Building occupancy			
2.	Interface to utilities			
3.	Interface to other systems			
4.	Monitoring non-HVAC systems			
5.	Building access to DDC			
6.	Building-specific conditions			
7.	Energy management programs			
8.	Alarm and message reporting			
9.	Advisories to local OWS			
10.	Building-specific items			
11.	Communications interfaces			
12.	Point mapping over the network			
E.	**Operator Related**			
1.	Operator advisories			
2.	Point definitions			
3.	Alarm limits			
4.	Alarm messages			
5.	Set points definitions			
6.	System diagnostics			
7.	Security/access			
8.	System start-up			
9.	Colorgraphics			
10.	Trends			
11.	Reports			
F.	**Maintenance Related**			
1.	DDC system functions			
2.	Network functions			
3.	Analog calibration			
4.	PID tuning			

161

Application Software—continued

Item	Description	Approved/ Not Approved	Not Submitted	Resubmit Date
F.	**Maintenance Related**			
5.	Software diagnostics			
6.	Network access			
7.	Equipment definition			
8.	Specific tuning and calibration issues			
9.	Equipment failsafe positions			
10.	Other job/system-specific items			
G.	**System's Management**			
1.	Standard/ad hoc reports			
2.	Report generation			
3.	Alarm history			
4.	Trend graphs			
5.	Interface to other management, graphic, and statistical software			
6.	Online interface with other computer systems			
7.	Data base management			
8.	Communications software			
9.	Software access			
10.	Customer programming			
11.	Other job/system-specific items			
H.	**Networking**			
1.	Interface software			
2.	Point mapping			
3.	Update/scan time			
4.	Use of third party data by the application software			
5.	Communications hardware definition			
6.	Network definition			
7.	Interface program/driver/communication protocol features			
8.	Software reliability checks for communication and data transfer			

Software Test Witnessing

Every DDC vendor has developed test procedures for testing and validating application software. The same is true for testing communications software and interfaces to third party systems. The commissioning agent should become familiar with the vendors' testing procedures and methods.

Witnessing of test procedures is an important QA measure. Some DDC vendors may consider it unnecessary to test the application software, and demonstrate it to the customer. Some vendors ship their controllers to the job site without testing the application software of individual controllers and OWS. Such conduct may speed up the hardware delivery to the job site, but most certainly will prolong system tuning, calibration, and start-up. It is much more convenient to troubleshoot software-related problems in the comfort of the vendor's office than to do it at the job site in the midst of construction activities.

A vendor's unwillingness to demonstrate the DDC software prior to shipment could be an indication of possible problems during system start-up. Remember, if it does not work on a test bench, it's not going to work in the real world either! The commissioning agent can use a modified application software checklist for test witnessing.

Review of Person Machine Interface (PMI)

With the advancement of color graphics software-operator interfaces have become graphical, animated, and display in color.

Graphical screens can offer a:

a. Static color graphics display of building and HVAC schematics
b. Dynamic presentation of analog readings, and the status of actual field points, bound to symbols on the graphics screen
c. Dynamic changes of shapes or colors of the depicted schematics based on actual states or modes of operation. Dynamic sections of the screen and the presented data are updated at periodic intervals to provide real time data presentation for the operators.

There are several graphics software packages on the market that provide DDC vendors with a broad selection of PMIs. Graphics software, especially for facilities with frequent upgrades, should be compatible with generic graphics packages (e.g. AutoCad). This could be an important time saving feature for facilities with ongoing construction activities, and frequent updates and modifications of PMI screens. Another reason for using colorgraphic software which resembles graphic software used for engineering drawing is that engineers, operators, and technicians are accustomed to a certain presentation of engineering drawings and controls schematics. Such presentations may not be as "flashy" as the three dimensional multi-color screens, but their value is in their simplicity, low cost of creating custom screens, and in their resemblance to drawings and sketches in the design documentation. However, drawings and control schematics can become cluttered with engineering information. Simplify such schematics to provide a clear presentation of the information needed for the systems operators.

Screens should be designed in a logical sequence (see discussion of tree structure that follows) allowing operators to move from one screen to another with a minimum of key strokes. Some DDC systems allow for custom design of the layout; other systems have "system maps" with a pre-determined order of graphic screens. For larger facilities with many graphic screens, the overall layout should be designed with future expansions in mind.

One of the more appropriate structures for graphics layout is a tree structure.

In the tree structure, if a facility were divided into geographical areas, the main menu would list:

• Main Area

• South Area

• North Area

• Other

For multiple functions, each area would then be divided into:

- BAS
- Utilities
- Facilities Management
- Energy Management
- BAS system layout and networking
- Other

In a networked environment, automation systems provide customized information for the "subscribers" on the network. For example, the BAS operator screen should take the operator to the Building menu of the selected geographical area. Then each building can have a menu of associated building systems.

The Utilities screens can present screens for individual heating and cooling plants and distribution systems. Each plant can have a menu of screens, for plant overview, chiller plant, boiler plant, energy, production, export, and so on. Further selections can take the operator to individual operating screens of systems associated with the chiller plant, boiler plant, and distribution systems.

Facilities Management screens should provide management information, such as overviews, management reports, trends, alarm reports, information on critical facilities, and other overviews from the entire facility.

Energy Management screens should provide energy information on consumption and demand for individual buildings, areas, intake points, and production, on daily, weekly, monthly, or annual bases. Such information can be formatted for distribution to individual building users to keep them informed and to encourage energy conservation.

BAS system screens should contain the network layout, network, systems, and controller, status, error rate, and so on.

Systems users and operators should have dedicated "hot keys," (hardware keys or "software keys" on the screens) which would allow them to move from one screen to another in the most efficient way. Efficiency, "friendliness" of the interface screens, and clear and concise format of presented data is the key for successful PMIs.

To create successful interfaces, DDC vendors or the systems integrator should seek input from the operators and other systems users. Because every system has its limitations, the interface layout should be structured and uniform, to provide maximum efficiency for all users.

Person Machine Interfaces

Item	Description	Approved/ Not Approved	Not Submitted	Resubmit Date
A.	**Overall Design**			
1.	Logical organization of screens			
2.	Main menus			
3.	Ease of moving from screen to screen			
4.	Data presentation			

Person Machine Interfaces—continued

Item	Description	Approved/ Not Approved	Not Submitted	Resubmit Date
B.	**Colorgraphics**			
1.	Static display			
2.	Point binding			
3.	Overviews			
4.	Building and/or equipment screens			
5.	Actual readings and set point displays			
6.	Equipment status display			
7.	Alarm status display			
8.	Details and control parameters			
9.	Control/change of values from the screens			
10.	Ease of moving from screen to screen			
11.	Dynamic update time			
12.	Use of hot keys			
C.	**Trending/ Report Generation/ Alarm Reporting**			
1.	History trends and graphs			
2.	Tuning graphs			
3.	Standard reports			
4.	Ad hoc reports			
5.	Alarm reports			
6.	Alarm messages			
7.	Operator advisories			
8.	Alarm history			
9.	History data files			
D.	**Systems Engineering & Management**			
1.	Graphics file transfer			
2.	History file sizes			
3.	Methods of file transfer to other media			
4.	OWS start-up/shut-down			
5.	System reboot			
6.	External storage media			
7.	Printer interface			
8.	Communications interface			
9.	Access codes			
10.	Operator log-on/log-off			

Controls Documentation Review

Controls and systems engineering documentation is important for future systems operations and maintenance. By the end of systems engineering, the controls vendor should have 90% of the documentation ready for review.

The following is a checklist for documentation review:

Controls Documentation

Item	Description	Approved/ Not Approved (Y/N)	Not Submitted	Resubmit Date
A.	**DDC Project Documentation**			
1.	System architecture			
2.	Control drawings			
3.	Field I/O list			
4.	Control cabinets layouts			
5.	Hardwired interlocks			
6.	Communications and interfaces			
7.	Wiring & terminations			
8.	Hardware cut sheets			
9.	Point definitions			
10.	Alarm definitions			
11.	Weekly scheduling			
12.	Set point definition			
13.	Loop definitions			
14.	Operating logic			
15.	Tables, calculations, etc.			
B.	**Test Results**			
1.	Factory testing			
2.	Calibration sheets			
C.	**Operating Manuals**			
D.	**Maintenance Manuals**			
E.	**Software Programs Manuals**			
F.	**Other Documentation**			

Factory Acceptance Testing

Factory acceptance testing of the DDC system should be conducted by the owner's commissioning agent at the vendor's site, prior to shipment of the system to the job site. This is a common practice for industrial automation systems, but seldom provided by building automation vendors. The reasoning is that the DDC systems are off-the-shelf systems, and each component is factory tested before it is shipped to the controls contractor or to the vendor's branch office.

However, system acceptance testing, as implied by the name, is testing of the entire system—the controllers, operator work stations, application software, and communications software and hardware. This should be functional testing of the application software, definitions, control and automation routines, point mapping over the network, testing of protocol services, communications with third party controllers, PMIs, and other application-specific issues.

It is often true that automation systems delivered to the job site without testing at vendor's premises (to meet deadlines for deliveries) have numerous start-up problems, and take much longer to calibrate and commission than do fac-tory tested systems. The seemingly good intention of delivering an untested sys-tem to meet construction deadlines may be a source of future problems during start-up, and can cause delays of the entire construction schedule. Systems tested prior to being shipped to the job site have far fewer problems during systems calibration and start-up.

The following is a sample checklist for factory acceptance testing:

System Acceptance Testing

Item	Description	Approved/ Not Approved (Y/N)	Not Tested	Retest Date
A.	**Software Definition on the Front-end and Downloaded Into Individual Field Controllers**			
1.	Point definition			
2.	Alarm definition			
3.	Set point definition			
4.	Software interlocks			
5.	Loop definition			
6.	Automation routines			
7.	PMI definition			
8.	Software download to controllers			
9.	Communication with controllers			
B.	**On Each Controller**			
1.	Hardware addresses			
2.	A/D conversion			
3.	Loop constants			
4.	Software definition			
5.	Hardware modules			
6.	Communication modules			
7.	CPU modules			
8.	Battery/charger			
9.	Redundancy			
10.	Spare I/Os			
11.	Cabinet spare capacity			
12.	Terminations strips			
13.	Grounding/spike protection			
14.	AC source/regulated power source			
15.	Transient protection			
16.	Labeling			
17.	Digital inputs			
18.	Digital outputs			
19.	Analog inputs			
20.	Analog outputs			
21.	Change of set point			
22.	Loop control			
23.	Demonstration of control logic			
24.	On-line point definition			

System Acceptance Testing—continued

Item	Description	Approved/ Not Approved (Y/N)	Not Tested	Retest Date
C.	**Communications/Protocols**			
1.	Protocol/driver services			
2.	Point mapping			
3.	Data transmission			
4.	Remote access			
5.	Remote reporting			
D.	**Operator Interfaces**			
1.	Log on/log off			
2.	PMI screens			
3.	Alarm reporting			
4.	Report generation			
5.	Trending			
6.	Alarm acknowledgement			
7.	Point control			
8.	Point reset			
9.	Change of set points			
10.	Change of alarm limits			
11.	Change of weekly scheduling			
12.	Data download/upload			
13.	Re-boot			
14.	Graphic file transfer			
15.	Data file transfer			
16.	Communications demonstration			
17.	Printer interfaces			

Identification of Customers

The Global Commissioning/TQM approach adds value to a process guided by traditional contractual obligations by identifying formal and informal customers. Formal customer relations are guided by contractual obligations. These do not change in the Global Commissioning/TQM approach, but are enhanced by teamwork, which creates informal customer relations. An example of formal customer relations is the HVAC design engineer who has a contractual obligation to provide a sequence of operation as part of the design documentation for the DDC systems engineer.

The Global Commissioning/TQM approach enhances the contractual obligations by promoting teamwork among the HVAC engineer, DDC systems engineer, and facilities engineer. Jointly, they develop application software which controls the HVAC system and automates the related building functions. Again, the key to successful design is in teamwork. This means pulling together the in-house and outside resources by identifying formal (contractual) and informal customers, and maintaining continuous quality assurance during systems engineering.

While the "external" (contractual) customer relationships remain unchanged, "internal" (informal) customer relationships are added to the systems engineering and review process. Involvement of in-house personnel provides the bridge between design and facility teams—so critical for successful automation systems operation and maintenance.

The commissioning agent should manage the process, and pull together the resources (called identification of customers in the Global Commissioning/TQM approach) for systems engineering and review tasks. Figure 6-2 could aid the commissioning agent in identifying customers for significant review stages of the systems engineering process. To what extent internal customers should be identified for any given job depends on the local circumstances, resources, and on the project itself.

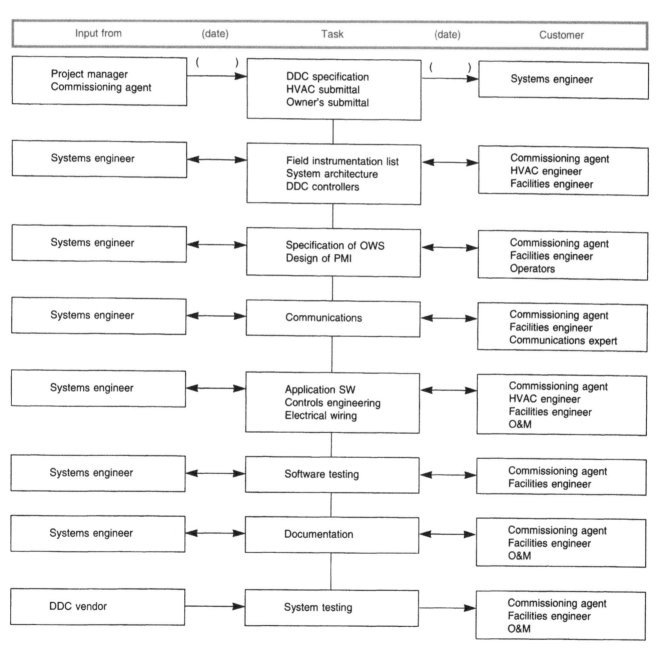

Figure 6-2. Customer identification chart

Figure 6-3 is an example of the commissioning agent's tasks at different stages of the systems engineering process.

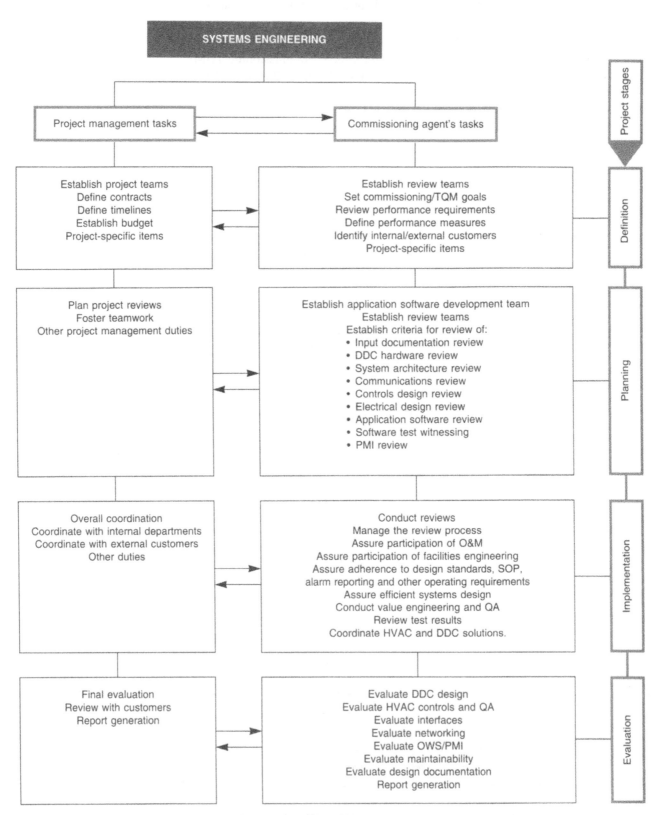

Figure 6-3. Commissioning agent's tasks at different stages of systems engineering

Chapter 7

DDC Installation Phase

Introduction

Installation of DDC systems is perhaps the phase best understood by project managers, construction managers, and contractors. Activities of DDC contractors during the installation phase of the project are similar to the activities of other contractors on site.

The systems engineering phase should be completed, and contractual obligations and responsibilities for related mechanical and electrical installations should be clarified prior to commencing the DDC installation phase. The following tasks should be addressed, defined, and scheduled for the installation phase:

1. Locating and visibly marking exact locations of all field instruments, control cabinets, and other hardware on the job site (by the DDC contractor)

2. Installation of hardware (by the mechanical and electrical contractors), as defined by the contract (the DDC vendor or contractor is responsible for the proper functioning of the DDC hardware installed by other contractors)

3. Installing electrical conduits and AC power wiring, providing wire labeling, testing, and termination on the DDC controller terminals, and installing of electrical field hardware, and AC distribution panels, as defined by the contract

4. Installing task lighting for DDC controllers

5. Installing controls and communications wiring and their termination on the DDC system

6. Installing communications components for interfaces and networks

7. Labeling and testing controls and communications wiring for the installation

8. Grounding DDC system

9. Installing controls hardware (by the controls contractor)

10. Marking, labeling, and color coding installed hardware

11. Installing DDC electronics into the installed control cabinets, and installing associated control systems hardware

12. Setting, calibrating, adjusting jumpers/switches, and tuning controls components

13. Downloading application software to individual DDC controllers

14. Setting up control loops

15. Providing initial field calibration

16. Installing operator interface hardware

17. Loading associated software to the OWS

18. Testing OWS and peripherals

19. Testing communications

20. Initial functional testing of controls and HVAC components in a standalone mode and from the OWS

21. Documenting as-built conditions relevant to controls and automation installation, and software and communications changes made during the DDC installation phase

22. Patching and painting (if contracted)

23. Removing existing controls hardware from the job site and their disposal (if contracted)

24. Cleaning-up the installation sites and mechanical rooms as requested by the construction management (e.g., for controls retrofit jobs)

Objectives for Installation Phase

The Global Commissioning/TQM approach set the following objectives for the DDC installation phase:

a. Conformance with site standards, specifications, and job documentation

b. Quality of installation, tuning, testing, and as built documentation

c. Periodic inspection of the installation by the owner's O&M personnel

These objectives provide two important results to the owner:

- Periodic Quality Assurance of the Installed DDC System
- Information Transfer From DDC Vendor to O&M Personnel

Quality Assurance of the Installed DDC System

Progressive reviews of the installed DDC system and of workmanship are part of a structured approach to quality assurance. The owner's commissioning agent responsible for progressive quality assurance should work closely with the controls contractor's project manager and construction manager to conduct periodic reviews and to achieve the desired quality measures.

Quality of the DDC installation depends on the quality of the contractors and their commitment to provide high quality workmanship. In real life, quality of DDC installations varies from one controls vendor or contractor to another. Some contractors are focused on quality; others may trade quality for meeting installation deadlines. Visual signs of poor workmanship in installed hardware, wiring, termination, and labeling are indicators pointing towards the contractor's lack of commitment to quality. Contractors who let their technicians get away with sloppy hardware installation will (most likely) have similar attitudes toward less visible tasks, such as instrument calibration, application software definition, and systems validation. Lack of systems engineering reviews and application software testing prior to shipment to the job site can also be a negative indicator, eventually resulting in systems problems and prolonged installation and systems start-up.

Complexity of DDC systems; coordination of hardware installation among the controls, mechanical, and electrical contractors; locations of field hardware, and their calibration; and possible modifications to the applications software, requires increased attention to as-built documentation. The controls contractor should make continuous corrections to the job documentation to capture all changes as they occur. Accurate DDC as built documentation is one of the most important tools for future operators and maintenance personnel.

The second, equally important objective for the commissioning agent is to develop a bridge for *transferring information from the DDC contractor to the operators and maintenance personnel.* This is accomplished by:

a. The owner's operators and controls technicians participating in the installation process

b. Transferring engineering documentation and O&M manuals to the owner

c. Adequately training the owner's operators and DDC technicians

While items (b) and (c) are part of the control contractor's contractual obligations, item (a) is not a common practice. Issues related to documentation and training are well understood and, usually, are part of the job contract. Participation of the owner's O&M personnel is outside of the formal contractual obligations. Interaction of O&M personnel with the DDC contractor's technicians provides a wealth of information for both parties. Most job coordination-related issues, or changes from the original design documentation, are discussed in job meetings. Participation of O&M personnel in job meetings may provide solutions to some installations related problems, and will enhance the O&M personnel's understanding of the built systems. Despite the obvious benefits, such participation is often underestimated by the DDC contractors, as well as by the owner's operations and maintenance management. Visitation of the job site by O&M personnel is often hindered by the daily load of maintenance problems. The daily workload may be a scheduling challenge the commissioning agent has to overcome. The commissioning agent also has to overcome the negative attitudes toward such visits by the participating contractors. To what level the operators and maintenance technicians should participate in the process, the frequency of their job site visits, and scheduling of these visits should be determined by the commissioning agent. Scheduling of job site activities should be coordinated with the construction manager, controls project manager and the owner's operations management.

In general, it is not customary in the building construction industry to have O&M personnel participate in the implementation process. In present practices, the project management, after job completion and prior to its turn-over to the owner, schedules a walk-through (job orientation) for the owner's O&M personnel. Walk-throughs for the mechanical trades, including controls, have been in practice in the construction industry for decades. They are beneficial for the familiarization of maintenance personnel with the DDC and HVAC hardware installation and their locations.

However, due to the complexity of building environmental control systems (HVAC and BAS and their interfaces), the traditional walk-throughs are insufficient for the HVAC and controls technicians. Additional time has to be dedicated for the controls technicians to learn DDC-specific items such as:

- Analog calibrations, switch and jumper settings

- Controls and automation sequences of operation (operating logic)

- Tuning of control loops and their testing

- Setting up communications and their testing

- Learning the application software loaded in individual controllers

- Systems maintenance and service procedures

- PMIs for the operators using laptop PCs, system access, software modifications, software upload/download and other systems as well as building operations procedures

Besides DDC-specific issues, O&M personnel should learn the operating and maintenance characteristics of the controlled mechanical systems, their performance characteristics, and shortcomings under normal and emergency operating conditions.

The expectation that the DDC operators and technicians could become proficient (and they need to be in order to operate and service a complex system), and absorb all relevant information during the walk-through without prior participation in the development process is unrealistic. Even the follow-up training and availability of good operating and maintenance documentation cannot fully substitute for real life exposure to systems development and implementation. The learning experience of the O&M personnel during the installation process is a very important component of their professional development. Observing the installation and being able to exchange information with the DDC technicians during installation and calibration is a prominent learning experience for the O&M personnel. No classroom training or simulation can substitute for such direct transfer of knowledge and experience.

Involvement of the O&M personnel in the periodic quality reviews of the installation provides not only continuous QA reviews, but also the on-site learning experience for operators and maintenance personnel.

However, unplanned and uncoordinated visits at the job site can be distracting and counterproductive. They can require extra time from the contractor's technicians, resulting in negative attitudes toward such visits. For this reason, the commissioning agent should schedule the site visits in coordination with the control contractor's project manager. For example, the DDC contractor may have some 100 VAV boxes to install, calibrate, and tune. The task could take several weeks to complete. The commissioning agent could schedule site visits spread over the duration of the VAV installation. A similar approach can be applied for setting up and tuning PID loops, and testing the operating logic, or mechanical systems start-up. Besides offering quality assurance, such visits provide a valuable learning experience for the O&M personnel.

The benefits of this on-site training approach will repay the cost many times over during regular systems operation, maintenance, and repair. An additional benefit is that the owner's technicians will also develop a feeling of ownership for the installed system.

Some controls contractors may have limited understanding and appreciation for O&M personnel involvement during the installation process. They may perceive them as a distraction that causes unnecessary delays. More seriously, they may be protective of their "trade" in fear of losing service contracts to the well-trained work force of the owner. The commissioning agent must develop a win-win atmosphere to overcome these prejudices, and develop a bridge between the DDC contractor and the owner's O&M personnel. Most DDC contractors understand that a successful DDC implementation means a satisfied customer, and that always means repeated and new business. Better trained O&M staff means fewer call backs, better utilization of the installed system, higher owner's satisfaction, and a better reference site to show to potential customers.

The project management and the commissioning agent should provide time and resources for O&M staff participation. Participation of O&M personnel in the process should be carefully planned, and their time on the job site should be optimized.

Performance Requirements and Measures

DDC vendors or contractors have written procedures for locating the controls hardware, hardware installations, wiring and wire terminations, systems grounding, field gear testing, and calibration. Similarly, they have procedures for loading and testing application software. The commissioning agent, along with the DDC vendor's or contractor's project manager, should review the contractor's procedures and develop a QA checklist relevant to the job.

The following is an example of a checklist with items found in most DDC installation jobs:

Checklist for DDC Installation

Item	Approved (Y/N)	Not Installed	Install by Date	Review Date	Notes
DDC Hardware Location					
Outside air sensor					
Room sensors					
Dust sensors					
Pipe sensors					
Control valves					
Dampers					
Meters					
DDC controllers					
Electrical AC cabinets/breakers					
Grounding systems					
Other job specific items					
Hardware Installed by Mechanical Contractor					
Thermal wells and taps					
Control valves					
Dampers and louvers					
In line meters					
Other job specific items					
Workmanship/Accessibility/Labeling					
Outside air sensor					
Room sensors					
Duct sensors					
Pipe sensors					
Actuators					
Electronics					
Electrical meters					
DDC Controllers					
Controls cabinets & their hardware					
External switches					
transformers					
relays					
Task lighting					
AC wiring					
termination					
testing					
labeling					

Checklist for DDC Installation—continued

Item	Approved (Y/N)	Not Installed	Install by Date	Review Date	Notes
Grounding termination					
testing					
labeling					
Controls wiring termination					
testing					
labeling					
Communications hardware					
Communications wiring termination					
testing					
labeling					
Other DDC hardware					
Initial Testing/Tuning					
Checking reference values and their calibration (voltages, ohms, milliamps, etc)					
Switch and jumper settings					
Application Software loading into local controllers					
Initial controls variables					
set points					
alarm values					
tuning of controls components					
Functional testing of input/output points					
PID loops					
Other job specific items					
Front End					
Setting up the front end					
Loading the front end software					
Loading the communications software					
Testing of front end					
Communications testing					
Controller software up/down load					
Point binding of color graphics					
Other job specific items					
Documentation					
Corrections for as built documentation					
Corrections of manuals					
Test, calibration sheets					

While the commissioning agent's main focus is on quality of the installed systems, the visiting O&M personnel should also provide feedback related to:

- Location of the installed hardware
- Accessibility for maintenance
- Visibility of markings and labeling
- Interfaces to existing systems
- Initial analog calibrations

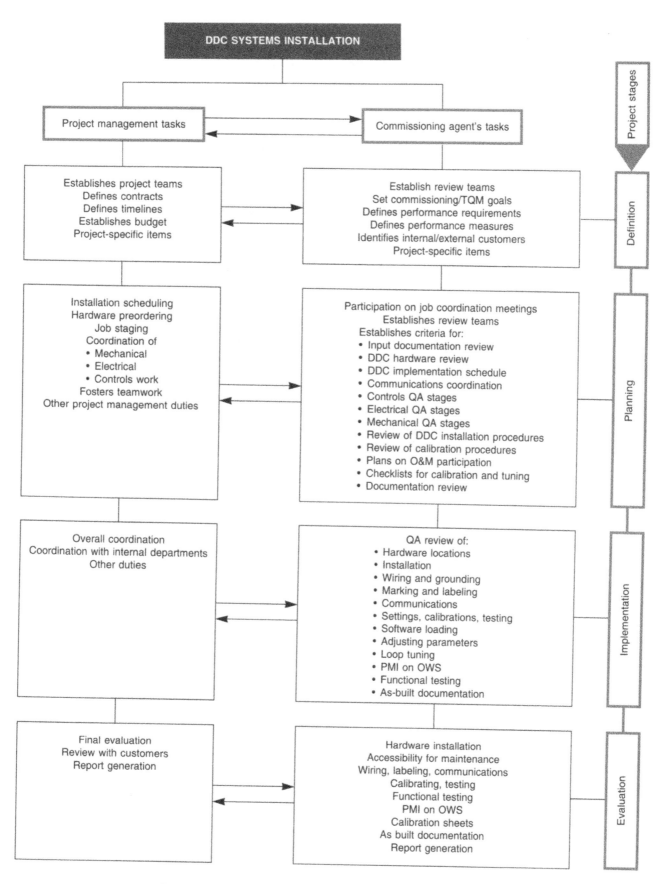

Figure 7-1. Commissioning agent's tasks for DDC systems installation

- Initial PID loop tuning parameters
- Existing communications systems
- System security
- Local access of the controllers and system
- PMI, reports, and other field and front-end related issues
- Systems operation under extreme and emergency conditions
- Start-up procedures.

Identification of Customers

Formal customer relationships on the job site are identified by contractual relations. Informal customers, such as O&M personnel, building occupants, facilities engineers, and so on, should be identified by the commissioning agent. There are both formal and informal customers with related tasks during the installation phase of the project. (Note: The commissioning agent should provide QA after completion of each task.)

Depending on the DDC system and the control contractor's procedures, there may be overlaps between systems engineering, installation, and commissioning phases of the project. Some vendors provide full software development, testing, and software installation during the systems engineering phase of the project. Other vendors provide extensive systems engineering assistance during the installation phase. From a project management and QA point of view, the first method is preferable. However, the latter can be equally successful if adequate resources and time are allocated.

Similar overlaps may occur for instrument calibration, set point variables, and alarm value definitions, definition of PID loop parameters, their tuning, and so on. There is often a very fine line between what can be done successfully on the test bench, and what has to be defined in the field, after the hardware installation. The commissioning agent should try to find such a balance to achieve a high quality of DDC installation while meeting the job time line.

Chapter 8

Installed Environmental Control Systems

Introduction

Commissioning installed environmental control systems is one of the most important phases of the design and implementation process in relation to systems performance. After months or years of effort, from the early stages of project definition and justification through design and construction, the completed job has to go through its final phase—*Commissioning and Quality Assurance* of the installed system prior to turning it over to the owner for standard operation and building occupancy.

Commissioning of environmental control systems (HVAC and DDC) is a complex task which should be done by professionals trained in the task.

For new construction jobs, HVAC commissioning is provided by firms specializing in Testing, Adjusting, and Balancing (TAB) hired by the general contractor, mechanical contractor, or the owner. DDC systems are commissioned by controls vendors or contractors' technicians.

For partial renovation jobs the situation of systems commissioning is less favorable. Many renovation projects are for specific building areas (floor, laboratory, department, HVAC unit), and are often carried out without prior air balancing surveys or recommissioning of the entire area upon completion of the job. This might create dangerous situations due to air contamination, especially in high risk areas such as manufacturing, laboratories, and health care facilities. Most renovations are done with limited scope and funding, and without adequate engineering assistance during the initial survey of existing buildings and mechanical systems. For "smaller" HVAC or floor renovation jobs, recommissioning of the renovated and affected adjacent areas is often omitted due to limited funding.

For new and retrofit jobs, commissioning of the HVAC and DDC systems is an independent undertaking with separate test reports submitted to the owner.

To improve the situation, the scope of commissioning should be determined by the commissioning agent in a specification of the scope of work. Base the decision as to the extent of air balancing surveys, TAB, and DDC commissioning for renovated and adjacent areas on the expected impact of renovation on the existing building and systems. The commissioning agent should develop procedures and coordinate the HVAC and DDC commissioning process in order to provide the owner with a unified report on the performance of the installed environmental control system.

There are several professional organizations that educate members, govern the HVAC commissioning process, and publish related procedures. Publications related to building systems commissioning are available to the engineering community. The most practical ones are the following:

- "Building Systems Commissioning" by the National Environmental Balancing Bureau (NEBB)

- "HVAC Systems Testing, Adjusting & Balancing" by the Sheet Metal and Air Conditioning Contractors National Association (SMACNA)
- "Guideline for Commissioning of HVAC Systems" from ASHRAE
- "Laboratory HVAC systems: Design, Validation, Commissioning" from the ASHRAE Technical Data Bulletin

Unfortunately, there are no comprehensive guidelines published for DDC systems commissioning. DDC systems are commissioned by the DDC vendors and contractors installing the systems. Reports submitted to the owner are often limited to calibration sheets. While these practices were sufficient for simple control systems, they are less satisfactory for modern facilities with their focus on systems performance and efficiency of operation.

General Objectives for Environmental Systems Commissioning

Building systems commissioning is well characterized in the NEBB publication "Building Systems Commissioning" as "...building systems commissioning is the process of providing the owner with a building that is complete, in compliance with the plans and specifications, and operationally and functionally ready to be taken over by the owner."

This includes:

a. Verifying system components under various conditions
b. Verifying interactions between systems and subsystems
c. Documenting system performance in reference to design criteria
d. Instructing operators how to operate the building systems and equipment.

These objectives should be adopted as a general guideline by building owners, commissioning agents, and organizations responsible for the implementation of environmental control systems.

Objectives for the Commissioning Agent

The commissioning agent should develop commissioning plans for all related environmental control systems affecting the safety and occupancy comfort of the related building. Execution of such a plan and coordination of commissioning activities between individual TAB and DDC technicians is the main responsibility of the commissioning agent.

The main tasks of the commissioning agent are:

- Review of individual commissioning plans and procedures of the TAB and DDC contractors, or development of a commissioning plan in absence of such plans.
- Coordination of individual commissioning activities on site
- Witnessing of commissioning activities and test procedures by assigned O&M personnel on site
- Verification of validity of test results, and assurance of compliance of environmental control systems performance with design criteria, perfor-

mance requirements, site specific requirements, and other governing standards

- Verification of commissioning reports prepared by the commissioning technicians, and their compilation to provide the owner with coordinated and uniform documentation
- Preparation of final commissioning documentation
- Preparation for final walk-through
- Instruction of operators and maintenance personnel on how to operate and maintain the installed systems
- Final turn-over of the job for full occupancy.

For retrofit installations the commissioning agent's tasks should include:

- TAB, controls and mechanical surveys of existing situations
- Re-balancing and re-commissioning of the renovated area, adjacent spaces and systems, or the affected building, in an effort to improve existing operating conditions.

The Commissioning Team

The team that commissions the environmental controls systems is led by the commissioning agent and should include:

- Appointed O&M Personnel
- TAB Technicians
- DDC Technicians

Cooperation among DDC and TAB technicians is of utmost importance in calibrating and balancing the entire environmental control system for optimum performance and efficient operation. While individual activities of TAB and DDC technicians commissioning the system are different at the beginning of the testing process, tuning the systems for optimum performance is a joint task. Participation of O&M personnel is important for witnessing and documenting the initial start-up parameters, and most importantly, for learning the systems' operating sequences for both normal systems operation and emergency conditions.

The commissioning agent should also establish a review team. The team could participate in witnessing the key commissioning tasks, reviewing commissioning reports, and approving deviations from the original design due to site- and system-specific conditions.

The commissioning review team should consist of the following professionals responsible for project implementation, as well as for safe and efficient building operation:

- Project Manager/Building Officials/Building Occupants
- HVAC Engineer
- Systems Engineer
- Facilities Engineer/Safety Technicians
- Systems Operators
- Maintenance Technicians

Performance Measures

Performance in the commissioning phase of the project can be measured in the following categories and by the specified individuals:

- Calibrating, tuning, and testing of DDC system, and testing and validation of the application software and controls algorithms
  (By the DDC vendors or contractors' technicians)

- Testing, Adjusting and Balancing (TAB) of HVAC systems
  (By TAB technicians or other balancing agents)

- Witnessing and validating test results and the environmental systems operation
  (By the commissioning agent and members of the commissioning team)

- Functional testing of the environmental control system under regular, extreme, and emergency operating conditions
  (Coordinated by the commissioning agent, and witnessed by the commissioning team)

- Planning level of training provided to the O&M personnel on how to operate and maintain the system under various conditions
  (By the project manager, and coordinated by the commissioning agent)

- Evaluating final performance of the installed system
  (By the commissioning agent to the owner)

It should be emphasized that the process is a coordinated effort aimed at providing the owner with a completed and functional environmental control system, which is a global approach in comparison to many existing practices of validating DDC installations independently from TAB activities on installed HVAC systems.

The following checklists are examples intended for use in the commissioning process. They do not substitute for testing procedures used by DDC contractors or certified TAB technicians. They should be used for verification of the HVAC and DDC system parameters and performance. The checklists are QA tools to assure that all systems are properly installed, checked out, and functional.

If properly modified for the specifics of the installed systems and site, they can aid in establishing a structured approach for O&M personnel to review the installed systems and their operating characteristics. Data collected on the checklists can be used for other facilities automation systems, for example as input data for Maintenance Management Systems. Nameplate and other field verified data collected during the commissioning process should be compiled in a format compatible with the Maintenance Management System. This can save on expenses in future field surveys.

Checklists and reports on test procedures, calibration, systems start-up parameters, and other data are essential for future reference. Collect and compile them in a format suitable for future reference for operating and maintenance departments. Such requirements sound like mundane tasks at the time of systems commissioning when there is the pressure to meet the job completion deadlines and turn over the building for occupancy. However, the long-term success of systems operation and maintenance depends very much on the quality of systems commissioning, training, and provided documentation. Operators and maintenance personnel utilize system commissioning and testing data in their reference library for many years after job completion and systems turn-over.

In an effort to standardize on figures, charts, and equations, and to narrow the information gap between the TAB, controls vendor's and owner's technicians, the most commonly used figures, charts, and equations are reprinted from national standards. They are part of the engineering tips following the checklists. For more comprehensive information on individual subjects, refer to relevant publications.

Checklist for Hot Water/Chill Water Systems

Item	HDW/SW	Device Model	End-to-End Accuracy (%)	Set Point Value	Alarm Limits Hi/Lo	Remarks
DDC Controller						
Application SW						
Other job specific items						
Hot Water						
Analog Inputs						
Hot Water Supply Temp						
Hot Water Return Temp						
Differential pressure						
Hot water flow						
Hot water valve feedback						
Hot water exchanger 1/3 steam valve feedback						
Hot water exchanger 2/3 steam valve feedback						
Hot water exchanger condensate temp						
Differential pressure valve feedback						
Pump variable speed						
Other job specific items						
Analog Outputs						
Hot water valve						
Hot water exchanger 1/3 steam valve						
Hot water exchanger 2/3 steam valve						
Differential pressure valve						
Pump variable speed drive						
Other job specific items						
Binary Inputs						
Hot water pump feedback						
Condensate receiver discharge						
Pump variable speed drive status						
Other job specific items						
Binary Outputs						
Hot water pump start/stop						
Other job specific items						

Checklist for Hot Water/Chill Water Systems—continued

Item	HDW/SW	Device Model	End-to-End Accuracy (%)	Set Point Value	Alarm Limits Hi/Lo	Remarks
Chill Water						
Analog Inputs						
Chill water supply temp						
Chill water return temp						
Chill water supply pressure						
Chill water return pressure						
Chill water flow						
Chill water valve feedback						
Pump variable speed						
Other job specific items						
ANALOG OUTPUTS						
Chill water valve						
Pump variable speed drive						
Other job specific items						
BINARY INPUTS						
Chill water pump feedback						
Pump variable speed drive status						
Other job specific items						
BINARY OUTPUTS						
Chill water pump start/stop						
Other job specific items						

Demonstration of Control Loops, Operating Sequences under Normal/Start-up and Emergency Conditions

Item	Operation Verified	Set points/limits Constants, etc	Documentation Verified	Re-test Date	Remarks
Hot Water					
Hot water exchanger temperature control					
Condensate receiver					
Hot water pump control					
Hot water dp control					
System start-up					
Emergency operation					
Operator advisories					
Other job specific features					
Chill Water					
Pump control					
Temperature/flow/pressure control					

Checklist for Hot Water/Chill Water Systems—continued

Demonstration of Control Loops, Operating Sequences under Normal/Start-up and Emergency Conditions

Item	Operation Verified	Set points/limits Constants, etc	Documentation Verified	Re-test Date	Remarks
System start-up					
Emergency operation					
Operator advisories					
Other job specific features					

Checklist for Communications/Point Mapping

Item	Building Controller	Application specific Controller	Serving AHUs	No. of Points	Communication Method/Protocol	Verified
DCC Controller						
Building						
Line integrity						
Terminations						
Transient protection						
EOL resistance/switch						
Communication speed						
Communication cards installation/setting						
Modems/hubs installation/setting						
Error rate						
Point verification*						
Command verification*						
DDC Controller						
Site Network						
LAN impedance						
LAN integrity						
Terminations						
Transient protection						
EOL resistance/switch						
Communication speed						
Communication cards installation/setting						
Modems/hubs installation/setting						
Error rate						
Point verification*						
Command verification*						

* For point verification use point definition printout

Notes: 1. For command verification use software logic

2. Point verifications at building controller level using a service PC; command verification at field point location.

Checklist for Air Handling Unit Control

Item	HDW/SW	Device Model	End-to-End Accuracy (%)	Set Point Valve	Alarm Limits Hi/Lo	Remarks
DDC controller						
Application SW						
Other job specific items						
Analog Inputs						
Outside air temp						
Preheat coil temp						
Mixed air temp						
Discharge air temp						
Return air temp						
Cooling coil temp						
Heating coil temp						
Static pressure						
Discharge air relative humidity						
Return air RH						
Supply fan variable speed drive control						
Return fan variable speed drive control						
Hot water supply temp						
Hot water return temp						
Chill water supply temp						
Chill water return temp						
Steam pressure						
Condensate temp						
Room temp						
Room temp						
Room temp						
Room temp						
Room RH						
Room RH						
Room RH						
Room RH						
Other job specific items						
Analog Feedback						
Damper						
Damper						
Damper						
Preheat valve						
Heating valve						
Cooling valve						
Humidifier						
Other job specific items						
Analog Outputs						
Damper actuator						
Damper actuator						

Checklist for Air Handling Unit Control—continued

Item	HDW/SW	Device Model	End-to-End Accuracy (%)	Set Point Valve	Alarm Limits Hi/Lo	Remarks
Analog Outputs						
Damper actuator						
Preheat coil valve						
Heating coil valve						
Cooling coil valve						
VSD output						
Humidifier valve						
Other job specific items						
Binary Inputs						
Binary input points						
Filter alarm						
Freeze alarm						
Supply fan status						
Smoke damper status						
Static pressure switch						
Condensate receiver pump status						
Other job specific items						
Binary Outputs						
Binary output points						
Supply fan start/stop						
Pump start/stop						
Other job specific items						

Demonstration of Control Loops, Operating Sequences under Normal/Start-up and Emergency Conditions

Item	Operation Verified	Set points/limits Constants, etc.	Documentation Verified	Re-test Date	Remarks
Mixed air control					
Outside air control/free cooling					
Preheat coil control					
Heating coil control					
Cooling coil control					
Dehumidification control					
Humidification control					
Air volume control					
Energy management functions					
System start-up					
Emergency operation					
Operator advisories					
Other job specific features					

Checklist for Terminal (VAV) Unit Control

Item	HDW/SW	Device Model	End-to-End Accuracy (%)	Set Point Value	Alarm Limits Hi/Lo	Remarks
VAV Controller						
Application SW						
Other job specific items						
Analog Inputs						
Room temp						
Supply air flow/pressure						
Supply air temp						
Other job specific items						
Analog Outputs						
Damper control						
Heating/cooling valve control						
Radiation valve control						
Other job specific items						
Binary Inputs						
Room switch/occupancy sensor						
Other job specific items						
Binary Outputs						
Fan start/stop						
Other job specific items						

Demonstration of Control Loops, Operating Sequences under Normal/Start-up and Emergency Conditions

Item	Operation Verified	Set points/limits Constants, etc.	Documentation Verified	Re-test Date	Remarks
Damper control					
Heating coil control					
Cooling coil control					
Air volume control					
Radiation heat control					
System start-up					
Emergency operation					
Operator advisories					
Other job specific features					

Checklist for DDC System Verification from Front-End OWS

Item	Approved Y/N	Not Ready	Re-commission Date	Remarks
Submittals				
DDC system manual				
Sequence of operation				
Control schematics				
System architecture				
Wiring diagrams				
Communications diagrams				
Point list				
Field controllers				
HDW Locations				
HDW product data sheets				
Analog calibration sheets				
PID loop check sheets				
Test reports on HVAC systems				
Test reports on DDC controllers				
Test reports on communications				
Applications software manual				
Alarm definition				
Set point definitions				
Program print-out				
Color graphics print-out				
Engineering manuals				
Operator's manuals				
Trouble shooting manuals				
Job specific operator's instructions				
Job specific troubleshooting instructions				
Other relevant instructions				
Verification				
Password/access definition				
Setting up the OWS				
SW loading to controllers				
SW loading to OWS				
Accessing point definitions				
Point modifications				
Accessing programs				
Program modifications				
Accessing color graphics				
Color graphics modifications				
Changing set points				
Changing alarm limits				
Operator commands				
Alarm report generation				
History report generation				
History trending				
Operator messages				
Time scheduling				

Checklist for DDC System Verification from Front-End OWS—continued

Item	Approved Y/N	Not Ready	Re-commission Date	Remarks
Verification				
Schedule modifications				
Demonstrations				
Point modification				
Point lock-out/disable				
Operation from color graphics				
Alarm acknowledgment				
Change of alarm limits				
Setting up reports				
Setting up history files				
Setting up history reports				
History file transfer to other media				
Setting up trends				
Change of time schedule				
Change of PID constants				
Setting up night setback				
Changing free cooling parameters				
Power failure restart				
Program on line modification				
Alarm message changes				
Other programming changes				
Demonstration of HVAC Functions (for each system)				
Time schedule				
Optimum start/stop				
VSD start-up				
Free cooling				
Hot water reset				
Set point calculations				
Loop cascading				
Related operating logic				
Operation under emergency conditions				
Demonstration of Interfaces				
Point display				
Point access				
Change of state reporting				
Alarm reporting				
Alarm acknowledgment				
Control commands				
Overrides				
Report generation				
DDC self diagnostics				
Communications network diagnostics				

190

Engineering Tips Related to Commissioning Checklists

Engineering tips are intended to provide basic technical information for project managers, and other non-systems engineering personnel. The information should aid in the understanding of the related subjects, testing results, and processes.

Engineering Tips for DDC Commissioning Checklists

Accuracy of control loops, and the DDC systems performance, depends on three components:

- Input points—from sensors and feedback devices
- DDC controller and the control logic
- Output points—to controlled devices

Input Points

Analog (AI):

- For temperature, humidity, pressure, etc., measurements
- Current (mA), or Voltage (VDC) input points

Digital (DI):

- Status, feedback, position, etc., indicators
- With Voltage or without Voltage on the contacts

Pulse (PI):

- Digital inputs, pulse counters, etc.

The following items will have an impact on the accuracy of Analog Input Points:

- Type of sensors used for the measured media
- Range for the actual measured variable
- Accuracy (at a point, over the entire range, etc.)
- Installation location, exposure, grounding
- Compatibility with the DDC system input module
- A/D, D/A converters
- Deviations from the calibration curve
- Sensor repeatability on rise/fall of measured variables
- Sensor dead band
- Drift over time
- Sensor linearity
- Stability over time
- Analog filter constant
- Accuracy of transmitters
- Accuracy of power source
- Error introduced by wire resistance, terminations, etc.

Besides analog input points, there are also analog data points (pseudo points), or software defined analog values with no associated field analog inputs. They are defined in the software only, and represent analog values from resulting calculations, tables, manual entries, and so on.

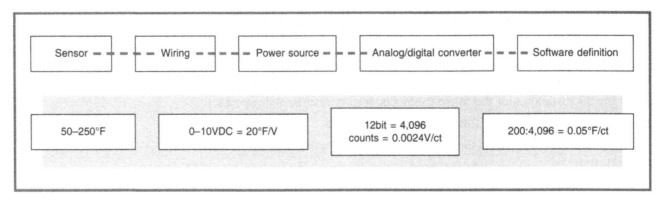

Figure 8-1. Software definition of analog sensor

Software Definition of an Analog Sensor

Figure 8-1 is an example of a software definition of an analog sensor (e.g., room temperature sensor). It aids in the understanding of definitions of individual components and their impact on end-to-end accuracy.

For example, an analog reading of a temperature sensor ranging 50-250°F has to be converted to an electrical value in volts (Figure 8-1). The voltage is then converted by an Analog to Digital (A/D) converter to digital counts in a 12 bit A/D converter which has 4,096 data elements. The A/D converted range of 0-10 VDC results in 0.0024 volts per count. The counts are then converted back to temperature per count (°F/count). In this example it is equal to 0.05°F/count, which is also called resolution of the sensor.

In the equation defined in the software,

$$Y = m(x) + b$$

the temperature is a function of counts (slope = °F/count), where

$$m = 0.05°F/Ct$$

and the intercept, when counts equal 0, b = 50°F.

The system or end-to-end accuracy is a function of software definitions, and the software calculations at a system's scan rates.

Software Definitions for Digital Input Points

Types of Digital Input Points:

- Contacts—Normally Open, or Normally Closed
- Switches, relays, pressure, differential pressure
- Sensors, thermostats
- Other contacts

Digital data points (pseudo points) are software-defined digital points which have no associated field hardware inputs. They are used in the software to simulate a digital input value in calculations and application logic.

The processor in the DDC computer interprets a Voltage value as "1" and a zero voltage value as "0." The programmer has to assign an associated status to 0s and 1s in the application software.

For example:
"1" = ON, OPEN, RUN, START, etc.
"0" = OFF, CLOSED, IDLE, STOP, etc.,
or any description (physical state) associated with the point status.

Note: Fire and safety detection devices should be hardware interlocked, with their secondary contacts reporting to the DDC system.

Output Points

Analog (AO):

- For damper and valve actuators or Electric to Pneumatic Transducers
- Current (mA), or Voltage (VDC) outputs

Digital (DO):

- Start/stop, two speed, re-position, etc.
- With Voltage, or without Voltage output

Pulse (PO):

- Pulse width modulation, etc.

The following items will have an impact on Analog Output Points:

- Valve, damper location
- Valve, damper characteristics/sizing
- Electrical, pneumatic actuation
- Actuator torque, travel, installation
- Actuator characteristics
- Accuracy under load/no-load conditions
- Actuator linearity
- Repeatability
- Resolution as a result of mechanical and DDC properties
- Positive feedback
- Failure mode—Normally Open/Normally Closed (NO/NC)
- Analog filter constant
- Analog signal (4-20mA, 0-10VDC, etc.)
- D/A, A/D converters
- Electric to pneumatic transducers

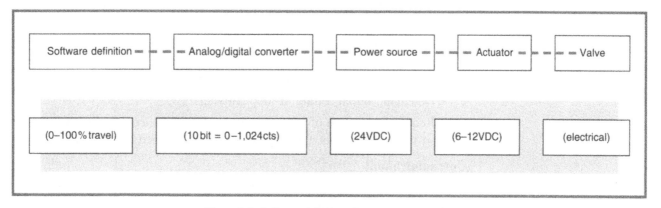

Figure 8-2. Software definition of analog output points

193

Software Definition of an Analog Output

The software definition of an analog output is similar to the definition for analog input.

For pneumatic actuation, the following elements are added to the example:

- Current to pressure transducer (i.e., 4-20 mA),
- Control air (i.e., 3-15 psi)
- Pneumatic valve (i.e., Open = 5 psi; Close = 10 psi)

Pulse outputs are digital solutions to analog outputs. The pulse output accuracy depends on a digital signal pulse (in milliseconds) scan rate, and the transducer.

The software definition impacts the final end-to-end analog accuracy, and the responsiveness of a control loop. Some analog points are results of calculations, or linearization, or are imported from tables and graphs. Some DDC systems require scaling of such data points, which means the programmer has to treat them as if they are analog hardware points. To scale them properly, the points have to be defined in the software for processing by the DDC controller's program.

The tuning of control loops is a complex task. This is due to:

- Definitions of PID constants
- Programming in the software
- Mechanical characteristics of actuators, valves, heating and cooling coils
- Characteristics of controlled media (steam, chill water, hot water, air), and their operating conditions.

This makes PID tuning a challenge for every job. No control loop can be fully tuned at the DDC vendor's bench without interfacing the DDC system to the real HVAC system.

The commissioning agent should focus on the following tasks, especially if the DDC technicians have difficulty tuning the control loops:

- Verify sensor locations
- Verify valve sizing, pressures, flows, and other design characteristics of a control valve
- Verify valve or damper control at full range, full stroke, full travel
- Verify loop stability under all operating conditions (worst conditions, high load conditions), and changing load conditions
- Verify testing, calibration, and software definition of all related components (span, transducers, positioners, spring ranges, stroking of actuators) under system load conditions
- Document initial setting of loop parameters
- Request correction of loops, mechanical systems, controls hardware, and software definitions, for loops that cannot be tuned (stabilized)

Communications

The commissioning agent should verify communication test reports, and request witnessing of systems communications from the building controllers or OWS, respectively. The commissioning agent should request written test results and verify the communication commissioning procedures from the DDC vendor or contractor, as well as from communications experts providing the means of communications.

The commissioning approach in communications and networking will become increasingly important due to the desire of the owners and DDC systems vendors to provide systems interoperability.

HVAC Systems Commissioning Checklists

DDC and HVAC systems are an integral part of the building environmental control systems. Even though the systems are designed and installed by different entities (HVAC and DDC contractors), upon completion of the job and their turn-over to the owner, they are operated and maintained by the same O&M department. From the safety and environmental comfort point of view, and from the building occupants' point of view, they are one and the same system. This view of the end user should also be adopted for the commissioning process. The commissioning phase of the project is the first real test of the integrity of the installed environmental control system. Although the HVAC TAB and the DDC system tuning and validation are done by independent organizations, the commissioning agent has to assure performance testing and tuning of the entire system. The commissioning agent and the review team should verify the functions of individual HVAC systems components controlled by the DDC system. This includes verification of the installed hardware (nameplates), their tested operating parameters, comparison with the design documentation, TAB results, DDC control logic, and the system's operation under normal and emergency operating conditions.

The following checklists help the commissioning agent focus on the commissioning of components of the HVAC system controlled by the DDC system.

Checklist for Fan and Fan Control

Item	Verification (Installed)	Notes
Nameplate Data		
Fan model #		
Serial #		
Class		
Type		
Manufacturer		
Motor volts/amps/kW		
Motor HP/rpm		
Model #		
Serial #		
Class		
Type		
Manufacturer		
Motor sheave make		
Fan sheave make		
Motor sheave diameter/bore		

Checklist for Fan and Fan Control—continued

Item	Verification (Installed)	Notes
Nameplate Data		
Fan sheave diameter/bore		
No. of belts/size		
Motor starter - amps		
Motor protection (heater)		
Constant/variable rpm		
Model #		
Serial #		
Class		
Type		
Manufacturer		
Variable Speed Drive		
Model #		
Serial #		
Class		
Type		
Options (PID and other)		
Manufacturer		

Test Data	Design	Actual	Calibrated	Remarks
Measured Data				
Speed (rpm)				
Volume (CFM)				
Static pressure (SP)				
Volts/frequency (V/Hz)				
KiloWatts/hour (kW/H)				
Phase 1,2,3 amps (A1/A2/A3)				
Reduced speed data at .... reduction:				
Speed (rpm)				
Volume (CFM)				
Static pressure (SP)				
Volts/frequency (V/Hz)				
Phase 1,2,3 amps (A1/A2/A3)				
KiloWatts/hour (kW/H)				
Reduced speed data at .... reduction:				
Speed (rpm)				
Volume (CFM)				
Static pressure (SP)				
Volts/frequency (V/Hz)				
Phase 1,2,3 amps (A1/A2/A3)				
KiloWatts/hour (kW/H)				
Reduced speed data at .... reduction:				
Speed (rpm)				

Checklist for Fan and Fan Control—continued

Test Data	Design	Actual	Calibrated	Remarks
Volume (CFM)				
Static pressure (SP)				
Volts/frequency (V/Hz)				
Phase 1,2,3 amps (A1/A2/A3)				
KiloWatts/hour (kW/H)				
DDC Verification				
Start/stop				
Name (acronym)				
DDC controller #				
Constant speed start/stop				
I/O module #				
Wire terminal #s				
Motor starter #				
Labeling				
Two speed start/stop				
I/O module #				
I/O module #				
Wire terminal #s				
Wire terminal #s				
Motor starter #				
Labeling				
Motor status (feedback)				
Field device				
DDC controller #				
I/O module #				
Wire terminal #s				
Labeling				
Variable speed drive				
DDC controller #				
Speed control I/O #				
Wire terminal #s				
Speed feedback I/O #				
Wire terminal #s				
VSD alarm I/O/ #				
Wire terminal #s				
Labeling				
Other I/Os				
Hardwired interlocks (name all)				
Labeling				
Software interlock (name all)				
Application programs:				
Time schedule				

Checklist for Fan and Fan Control—continued

Test Data	Design	Actual	Calibrated	Remarks
Economizer cycle				
Other programs				
Prerequisite to start/stop				
Alarm reporting verification				
Witnessed at OWS & DDC controller				
Maintainability				
Access				
Maintenance instructions				
Lighting				
Other associated items				

Engineering Tips Related to Fan and Fan Control Commissioning Checklist

Motor Speed—revolutions per minute (rpm), measured by tachometers, odometers, stroboscopes, etc., Requested tolerance: $+/-1$ to $+/-2$ %.

Fan Static Pressure (SP)—the difference between the total fan pressure at the intake and the static pressure at the fan outlet

$$\text{Fan SP} = \text{SPout} - \text{total P (in. w.g.)}$$

Fan Characteristics—each fan has its associated fan characteristic curves. The examples in Figures 8-4 to 8-7 are for different types of fans.

Fan Classes—developed by Air Movement and Control Association, Inc. for classification of fan structural limitations (Fig. 8-8).

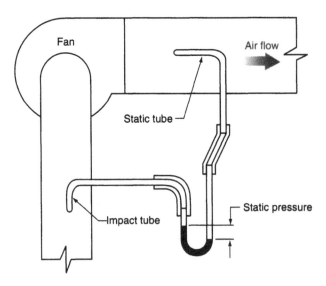

Figure 8-3. Fan static pressure (adapted from ASHRAE III, fig 15)

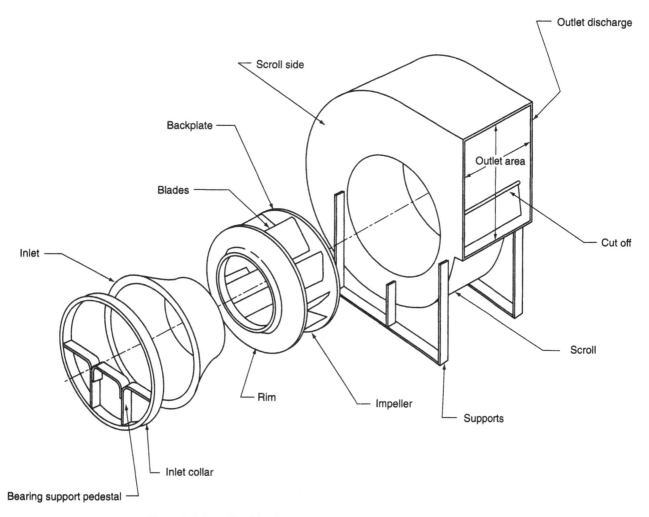

Figure 8-4. Centrifugal fan (adapted from SMACNA 5.1, figs 5-1 and 5-2)

Fan Laws—used to determine fan performance under different conditions

$$\frac{Q_2}{Q_1} = \frac{RPM_2}{RPM_1} \;......\text{where Q = airflow [cfm]}$$

$$\frac{P_2}{P_1} = \left(\frac{RPM_2}{RPM_1}\right)^2 \;......\text{where P = pressure (in. w.g.)}$$

$$\frac{FP_2}{FP_1} = \left(\frac{RPM_2}{RPM_1}\right)^3 \;......\text{where FP = fan brake power in HP}$$

$$\frac{d_2}{d_1} = \left(\frac{RPM_2}{RPM_1}\right)^2 \;......\text{where d = air density in lb/ft}^3$$
$$\text{(standard air d = .075 lb/ft}^3)$$

The air density has no effect on the airflow. However, the fan brake power and the pressure will vary with the density.

$$\frac{FP_2}{FP_1} = \left(\frac{RPM_2}{RPM_1}\right)^2 \times \frac{d_2}{d_1}, \text{ and}$$

$$\frac{P_2}{P_1} = \left(\frac{RPM_2}{RPM_1}\right)^2 \times \frac{d_2}{d_1}$$

199

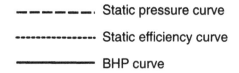

- — — — — — Static pressure curve
- ················· Static efficiency curve
- ———————— BHP curve

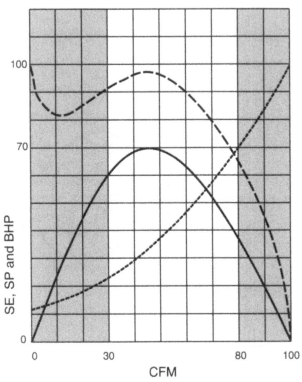

Figure 8-5. Forward curved centrifugal fan curve

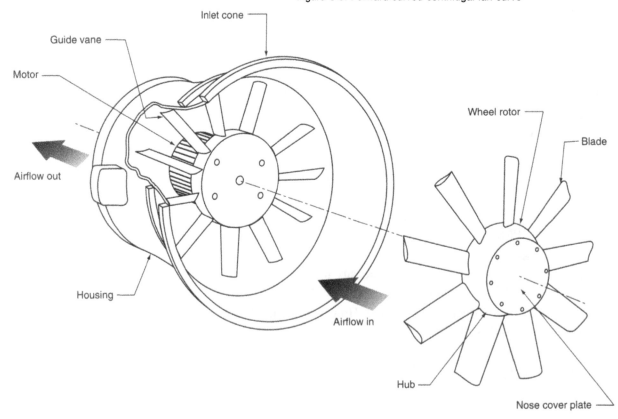

Figure 8-6. Axial fan (adapted from SMACNA 5.2, fig 5-5)

— — — — — — . Static pressure curve

· · · · · · · · · · · · · · · Static efficiency curve

———————— BHP curve

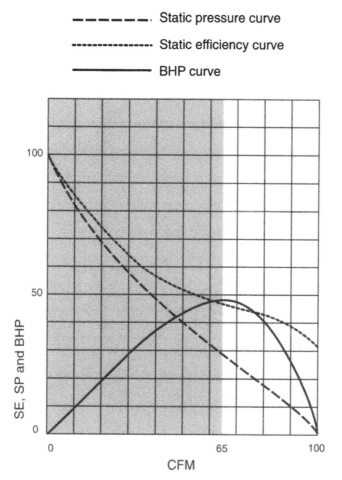

Figure 8-7. Propeller axial fan curve (adapted from SMACNA 5.3, fig 5-6)

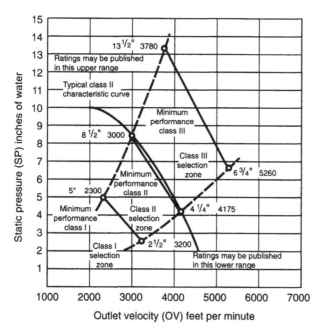

Figure 8-8. Fan class standards (adapted from SMACNA 5.4, fig 5-10)

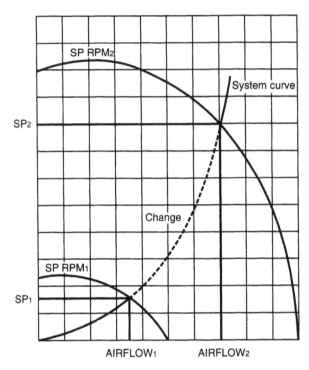

Figure 8-9. Effect of rpm change on static pressure (adapted from SMACNA, 5.12, fig 5-25)

Understanding these equations is important because of the advancement of Variable Speed Drives (VSD) utilized for Variable Air Volume (VAV) systems as well as for Variable Exhaust Systems (VES). In both cases the system resistance (louvers, dampers, filters, coils, ductwork, diffusers, etc.) remains the same. However, the change of rpm causes changes of airflow and change of pressure, as shown in Figure 8-9.

The equation for the Fan Brake Horsepower of a single phase motor is:

$$BP = Amps \times Volts \times Power\ Factor \times Efficiency/746$$

The equation for the Fan Brake Horsepower of a three phase motor is:

$$BP = Amps \times Volts \times Power\ Factor \times Efficiency \times 1.73/746$$

Engineering Tips Related to the Damper and Damper Control Checklist

Air handling unit volumetric dampers are multiblade parallel or opposed action dampers controlled by damper actuators. Dampers can be connected by linkages, controlled by one or more actuators to open/closed positions, or modulated to maintain desired air volume. Outside air dampers can have a minimum position or dedicated minimum air dampers; outside, return and exhaust dampers should be balanced by a TAB technician to deliver the desired air ratio.

Dampers are controlled by analog or digital output(s) from the DDC; minimum air dampers are usually interlocked to the supply fan. Since most air handling units do not have volumetric controls, it is desirable to have a positive damper feedback (potentiometer) of the damper position.

Checklist for Damper and Damper Control

Items	Verification	Notes
Nameplate Data		
Model #		
Serial #		
Type		
Manufacturer		
Damper actuator model #		
Serial #		
Type		
Manufacturer		

Test Data

Item	Design	Actual	Calibrated	Remarks
Measured Data				
Number of dampers				
Linkage setting				
Actual flow				
Actual velocity				
Minimum position setting				
Air leakage at closed position				
Actuation time 0-100%				
DDC Verification				
Name (acronym)				
DDC controller #				
I/O module #				
Wire terminal #s				
Actuator el/pneum				
Voltage/mA/psi				
Torque				
Adjustments marked at linkages				
Minimum position				
Labeling				
Controlling point				
Name (acronym)				
DDC controller #				
I/O module #				
Wire terminal #s				
Status (feedback)				
Field device				
I/O module #				
Wire terminal #s				
Labeling				
Hardwired interlock (name all)				
Labeling				

Checklist for Damper and Damper Control—continued

Item	Design	Actual	Calibrated	Remarks
Application program				
Sequence of operation				
PID control(s)				
Control set points				
Fall back position				
Alarm reporting				
Other associated items				
Witnessed at OWS & DDC controller				
Maintainability				
Access				
Maintenance instructions				
Lighting				

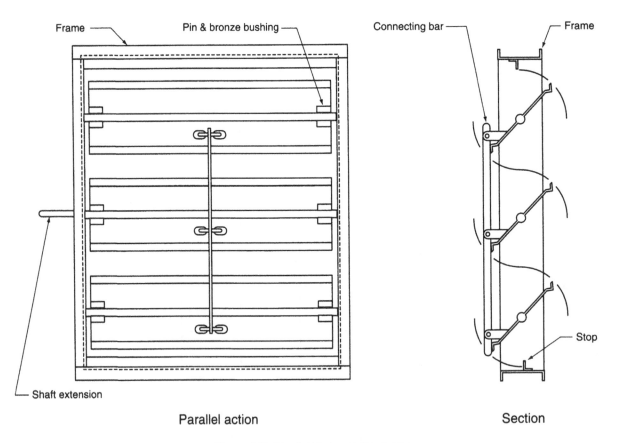

Parallel action

Section

Figure 8-10. Parallel blade volume damper

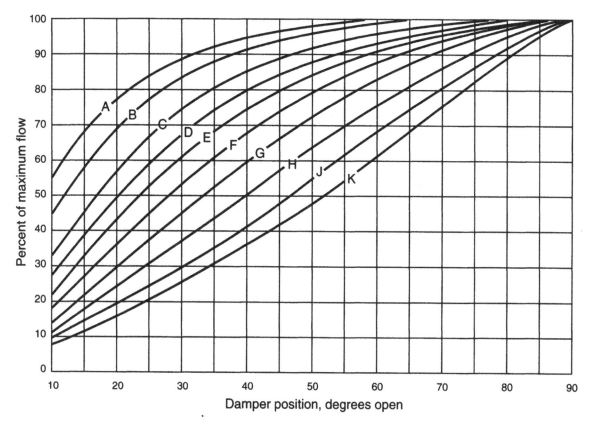

Figure 8-11. Flow characteristics

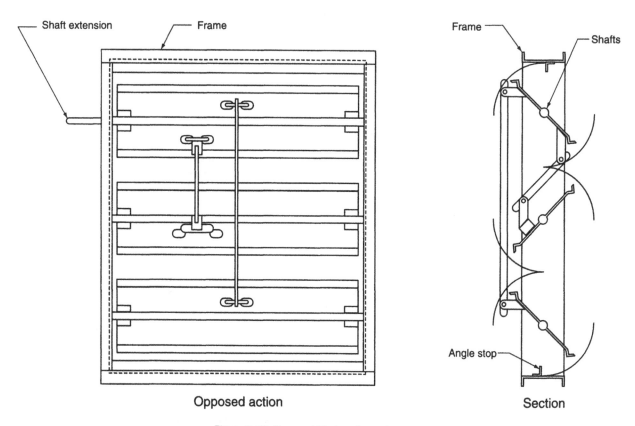

Opposed action

Section

Figure 8-12. Opposed blade volume dampers

205

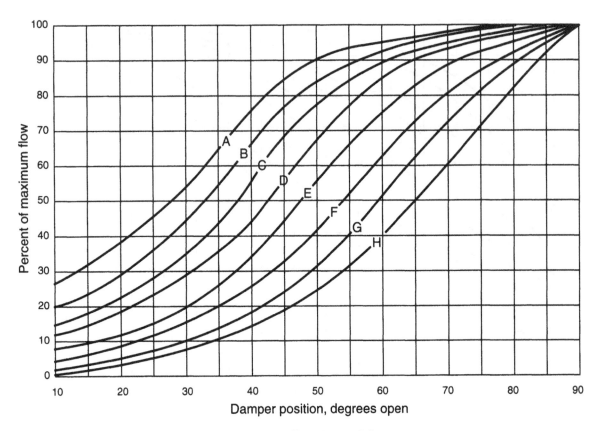

Figure 8-13. Flow characteristics

Parallel-leaf dampers		Opposed-leaf dampers	
Open damper resistance, percent of system resistance	Flow characteristic curve	Open damper resistance, percent of system resistance	Flow characteristic curve
0.5 – 1.0	A	0.3 – 0.5	A
1.0 – 1.5	B	0.5 – 0.8	B
1.5 – 2.5	C	0.8 – 1.5	C
2.5 – 3.5	D	1.5 – 2.5	D
3.5 – 5.5	E	2.5 – 5.5	E
5.5 – 9.0	F	5.5 – 13.5	F
9.0 – 15.0	G	13.5 – 25.5	G
15.0 – 20.0	H	25.5 – 37.5	H
20.0 – 30.0	I		
30.0 – 50.0	J		

Exhibit 8-14. Ratio of open damper resistance to system resistance flow characteristics curves (reprinted from SMACNA pg. 6.6, table 6.1)

Percentage of air flow at different openings via the damper depends on the percentage of damper resistance to the entire air systems resistance (see Figure 8-14)

Engineering Tips Related to Filter Alarms Checklist

It is a common practice to provide a DDC system report for "dirty filter" alarms. The filter differential pressure switch should be set at a value that does not cause nuisance alarms yet reports a pressure drop across the filters to indicate the need to clean or change the monitored filter. Under known operating conditions and with a scheduled, periodic filter replacement program, it may not be necessary to have filter alarms reported via the DDC system.

Checklist for Filter and Filter Alarm

Items	Verification	Notes		
Nameplate Data				
Model #				
Serial #				
Type				
Manufacturer				

Test Data				
Item	Design	Actual	Calibrated	Remarks
Measured Data				
Number of filters				
Actual flow				
Actual velocity				
Differential pressure				
DDC Verification				
Filter alarm				
Name (acronym)				
DDC controller #				
Field device				
I/O module #				
Wire terminal #s				
Labeling				
Software name				
Alarm message				
Witnessed at OWS & DDC controller				
Maintainability				
Access				
Maintenance instructions				
Lighting				

Checklist for Coils and Coil Control

Items	Verification	Notes
Steam Coils		
Nameplate Data		
Model #		
Serial #		
Type		
Manufacturer		
Number of coil sections		
Steam pressure		
Coil capacity		
Coil temperature dp		
Steam trap nameplate data		
Size		
Model #		
Serial #		
Type		
Manufacturer		
Coil vent		
Size		
Model #		
Serial #		
Type		
Manufacturer		
Control valve/size		
Model #		
Serial #		
Type		
Manufacturer		
Control valve actuator/torque		
Model #		
Serial #		
Type		
Manufacturer		
Condensate return piping		
Strainers, drip legs, etc.		
Condensate receiver		
Liquid mover		
Water Coils: Hot Water/ Chill Water		
Nameplate Data		
Coil model #		
Serial #		
Type		
Manufacturer		

Checklist for Coils and Coil Control—continued

Items	Verification	Notes
Water Coils: Hot Water/ Chill Water		
Nameplate Data		
Number of coil sections		
Coil capacity		
Coil temperature/dp		
Pump		
Model #		
Serial #		
Type		
Manufacturer		
Pump rpm		
Pump impeller		
Control valve/size		
Model #		
Serial #		
Type		
Manufacturer		
Control valve actuator/torque		
Model #		
Serial #		
Type		
Manufacturer		
DX Coil		
Nameplate Data		
Coil model #		
Serial #		
Type		
Manufacturer		
Number of coil sections		
Coil capacity		
Coil temperature/dp		
Pump		
Model #		
Serial #		
Type		
Manufacturer		
Pump rpm		
Pump impeller		
Control valve/size		
Model #		
Serial #		
Type		

Checklist for Coils and Coil Control—continued

Items	Verification	Notes
DX Coil		
Nameplate Data		
Manufacturer		
Control valve actuator/torque		
Model #		
Serial #		
Type		
Manufacturer		

Test Data

Item	Design	Actual	Calibrated	Remarks
Measured Data				
Pressure				
Valve dp				
Coil air flow				
Coil differential pressure				
Coil differential temperature				
DDC Verification				
Name (acronym)				
DDC controller #				
I/O module #				
Wire terminal #s				
Actuator				
Voltage/mA/psi				
Torque				
Fall back position				
Labeling				
Status (feedback)				
Field device				
I/O module #				
Wire terminal #s				
Labeling				
Hardwired interlock (name all)				
Labeling				
Application software				
Control inputs (name all)				
Sequence of operation				
PID loop constants				
Control set point				
Calculations				
Alarm reporting				

Checklist for Coils and Coil Control—continued

Test Data—continued

Item	Design	Actual	Calibrated	Remarks
DDC Verification				
Other associated items				
Witnessed at OWS & DDC controller				
Maintainability				
Accessibility				

Engineering Tips Related to Coils and Coil Control Checklist

Every HVAC system, with the exception of ventilation systems, utilizes some form of heat exchange to transfer heat from one media to another. In air handling units, coils are used to preheat, reheat, cool, and dehumidify the air to achieve desired parameters (temperature and relative humidity) for the building. In most coils, heat transfer takes place between hydronic systems (or steam) and air systems. The process is controlled by flow variations and heat transfer rates.

Figure 8-16 is a chart from the 1991 ASHRAE Application Handbook, and depicts the effect of supply water flow rate change on the percent of heat transfer for a 20°F temperature drop (deltaT) in the coils. As shown on the chart, to reduce the heat output to 50%, the heating valve must be commanded to allow 10% of the designed flow rate [that may be as much as 90% travel (rate of flow) for the stem of a fully opened valve].

For most heating coils utilized in air handling units, the flow rate change is not the only controlled variable affecting the heat transfer. The temperature of

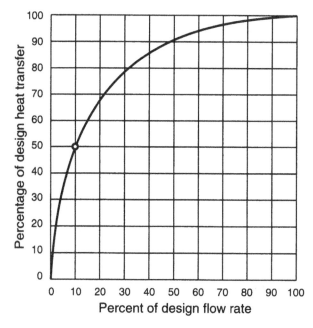

Figure 8-15. Effects of flow rates on heat transfer

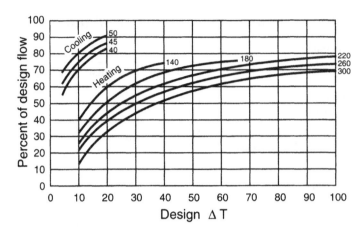

Figure 8-16. Design flow vs. design differential temperature for 90% terminals heat transfer (adapted from 1991 ASHRAE Handbook, 34.7, fig. 2)

supplied hot water has a profound effect on the percentage of designed flow rate and, consequently, on the percentage of the heat transfer change.

As Figure 8-16 demonstrates, for a 20°F deltaT coil, 140°F hot water requires a 60% design flow, while at a 220°F hot water supply temperature, it requires only 44% to maintain a 90% coil heat transfer. Note that the design limit for low pressure hot water is 250°F.

Figures 8-15 and 8-16 show the importance of proper tuning of associated PID loops—the loop controlling the hot water exchanger supply temperature, and the loop controlling the coil's discharge air temperature—to achieve accurate control, and optimum systems operation.

On the upper left corner of Figure 8-16, the chill water curves show the effects of different chill water supply temperatures on the designed flow to maintain a 90% coil heat transfer.

A generalized chill water coil chart (Figure 8-17) shows the heat transfer for a typical ARI rated coil for a 45°F supply water, a 10°F deltaT coil, and 80°F

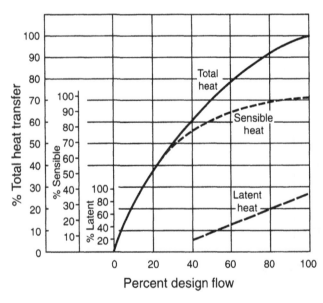

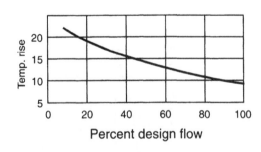

Figure 8-17. Chill water flow vs. temperature rise and heat transfer (adapted from 1991 ASHRAE Handbook, 34.7, fig. 3)

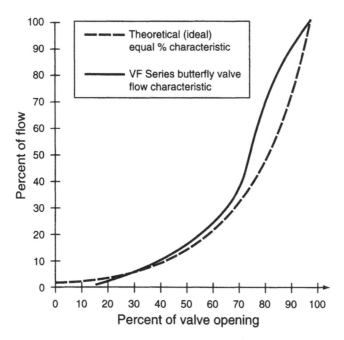

Figure 8-18. Butterfly valve flow characteristics (Johnson Controls, Inc.)

dry bulb/67°F wet bulb entering air temperatures. (For more accurate representation refer to coil manufacturers charts.)

The flow curves are influenced by the entering water and air temperatures, flow rates, and the coil itself. Changes of these factors influence the shape of the curves. However, the curves remain non-linear. The non-linearity of a coil is linearized by a valve flow characteristic. Figure 8-18 shows butterfly valve flow characteristics.

The relatively linear characteristic of a combined coil and equal percentage valve characteristic is shown in Figure 8-19.

There is a difference between hot or chill water coil characteristics and steam coil characteristics. Steam coil characteristics are influenced by the entering air temperature, steam supply pressure, pressure drop, and the coil characteristics itself. Control of the heat output of a steam coil is accomplished by controlling the steam flow rate and, consequently, the condensation rate. By closing a steam valve, the flow rate is reduced. However, the condensation in the coil continues at the same rate. This causes a decrease of valve outlet pressure (increase of pressure drop across the valve) and an increase in steam velocity up to the critical pressure drop (about 45% of supply pressure), from which point the steam velocity remains unchanged. A valve with equal percentage characteristics is suitable for proportional control of a steam coil (Figure 8-20).

It is important to understand these concepts in order to understand and evaluate the results of coil testing and commissioning. Another important factor is to understand the temperature distribution measured at the discharged air side at the face of the coil. While the average temperature of a hot water coil is fairly representative for the entire coil, temperature differences of large steam coils can be significant because of the rate of steam condensation in the coil. The steam coil discharge air temperature, if measured by one averaging temperature element, can be a misrepresentation of the coil temperature distribution. For example, a preheat coil temperature reading

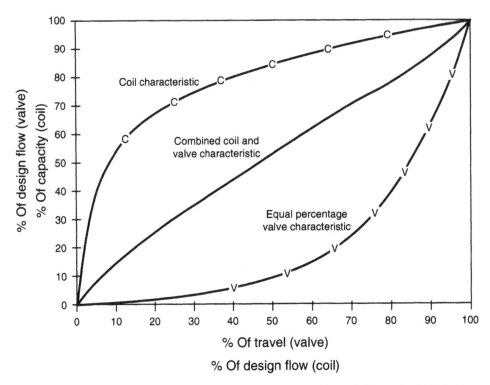

Figure 8-19. Combine coil and equal percentage valve characteristics (Johnson Controls, Inc.)

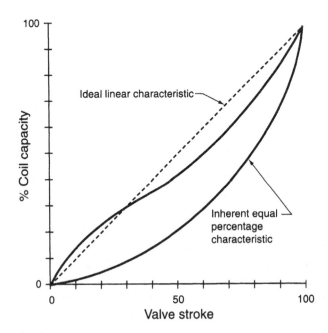

Figure 8-20. Inherent equal percentage valve characteristics due to changing differential pressure at the valve (Johnson Controls, Inc.)

might show temperatures in the 50's (measured by the averaging temperature element), yet the freeze stat set at 38 degrees will report freeze conditions due to low temperatures. Two inches of the freeze stat element exposed to the set temperature (38°F) will trip the freeze stat.

Preheat coils which are not properly sized, drained, and vented, and directly exposed to below freezing air temperatures, are in danger of coil freeze-ups regardless of their DDC controls.

Control Valves and Actuators

Control valves control the flow of hot water, chill water, and other liquid or steam into the coils. The flow regulation is accomplished by stroking (opening/closing) the valve, thus introducing resistance, pressure change, and consequently flow change of the heating or cooling media. Proper selection and sizing of control valves is an important factor for efficient systems operation. Oversized valves cause cycling (hunting) and overall poor control under partial load conditions; undersized valves might not be sufficient to provide adequate cooling or heating capacity, and can impose increased resistance that has to be overcome by pumping capacity. Another problem is related to uncontrollable flow rates occurring at a low opening (10-20%) of the valve travel. To calculate the uncontrollable flow rate, the engineer has to find the ratio between "maximum flow" and "minimum controllable flow" for the designed pipe, valve style, and pressure differential.

Valve Styles

There are two valve styles used in HVAC applications: butterfly and globe.

- Butterfly valves are used as two position valves (on/off), or as mixing valves in chill water systems.
- Globe valves are used in proportional control applications.

The generic equation for water flow is:

$$\text{GPM} = \text{Valve flow rate coefficient} \times \sqrt{\text{DP}}$$

The equation for steam flow in lbs of steam/hr is:

Lbs/hr = Valve flow rate coefficient × (0.41) Absolute inlet pressure
for steam below 10psig, or × (1.76)
for steam above 10psig.

Based on the flow rate and differential pressure, the valve flow coefficient can be calculated as:

$$\text{Cv} = \frac{Q}{\sqrt{\text{DP}}}$$

The valve sizes and types can be found for the calculated values in the valve manufacturer's tables.

Control valves are controlled by either pneumatic or electrical actuators. The most important aspect related to actuators is the force necessary to overcome the dynamic pressure of the hydronic system when moving the plug in the valve to regulate flow. Another important factor is the time of valve stem travel from 0-100% opening. The time factor has to be matched in the software definition. For actuator sizing refer to manufacturer's tables for the given valves.

Checklist for Air Distribution Systems and Control

Test Data	Design	Actual	Calibrated	Remarks
Measured Data				
Outside air CFM/%				
Minimum air CFM/%				
Preheat coil dT				
Cooling coil dT				
Reheat coil dT				
Discharge duct size/area				
Discharge air CFM/pressure				
Discharge duct velocity				
Discharge duct static pressure				
Return duct size/area				
Return air CFM/pressure				
Return duct velocity				
Return duct static pressure				
Mixing box				
Mixed air CFM/pressure				
Mixed air static pressure				
Exhaust duct size/area				
Exhaust air CFM/pressure				
Exhaust duct velocity				
DDC control witnessed				
TAB settings marked				
Test reports reviewed				
Terminal box #:				
Area served				
Make/model				
Type/size				
Effective area				
Supply CFM/fpm				
Damper actuation				
Valve actuation				
Coil dT				
Supply/exhaust ratio cfm/dp				
Interlocks				
Noise level				
Space temp/air change				
DDC control witnessed				
TAB settings marked				
Test reports reviewed				

Engineering Tips Related to Air Distribution Systems and Control Checklist

Most commonly used air systems are single duct and multi-duct (dual duct) systems. Air systems can distribute constant volume or variable volume air into their designated areas, usually zones. At the zone terminal units, the environment (zone, room, area) can be controlled by controlling the air volume, reheating, cooling, or humidifying of the supply air. Zone unit controls can be interlocked with perimeter radiation heating or exhaust systems to control the desired conditions within designed parameters. Proper balancing of air systems is essential to maintain the designated ratio between supply and exhausted air, room pressurization, adequate ventilation, temperature, and relative humidity. Constant pressure systems (single or dual duct) can be retrofitted to variable volume systems by controlling the duct pressure (by controlling the supply air), and converting the terminal units to variable air volume units (VAV). By conversion of the supply air to VAV, the exhausted air should be matched to maintain pressurization requirements throughout the building. To assure economical systems operation, mechanical heating and cooling should be utilized only if the outside air conditions cannot provide adequate heating or cooling. Temperature, relative humidity, and air volume (CFM) setback should be programmed for unoccupied times.

The duct distribution system as well as terminal systems should be sized and balanced to deliver designed (calculated) air volumes at designed velocity. Adequate heat and humidity control should be accompanied by comfortable air velocity and acceptable noise levels. Room air velocity of 50 feet per minute (fpm) is acceptable for most occupants. The range for occupancy comfort is between 20-70 fpm. Acceptable noise levels for different rooms are listed in the related annual ASHRAE handbooks.

Checklist for Rooftop Units/Heat Pumps and Control

Items	Verification	Notes
NAME PLATE DATA		
Unit make/model		
Type/size		
Filters type/size		
Fan sheave		
Diameter/bore		
No of belts/size		
Motor make/frame		
HP/rpm		
V/A/kW		
Fan type/size		
Evaporator make/model		
Type/size		
Condenser		
Refrigerant		
Compressor model/type		

TEST DATA

Item	Design	Actual	Calibrated	Remarks
MEASURED DATA				
Total air flow				
Total static pressure				
Suction static pressure				
Discharge static pressure				
Discharge air flow				

Checklist for Rooftop Units/Heat Pumps and Control—contd.

TEST DATA—continued

Item	Design	Actual	Calibrated	Remarks
MEASURED DATA				
Discharge DB/WB				
Return static pressure				
Return air flow				
Return DB/WB				
Fan HP/rpm				
V/A/kW				
Condenser suction pressure/temp				
Condenser pressure/temp				
Compressor HP/rpm				
Compressor V/A/kW				
Lo/Hi pressure cut-out				
Condenser fan HP/RPM				
Airflow				
V/A/kW				
DDC Verification				
Name (acronym)				
DDC controller #				
I/O module #				
Wire terminal #s				
Labeling				
Status (feedback)				
Field device				
I/O module #				
Wire terminal #s				
Labeling				
Hardwired interlock (name all)				
Labeling				
Application software				
Control inputs (name all)				
Sequence of operation				
PID loop constants				
Control set point				
Calculations				
Alarm reporting				
Other associated items				
Witnessed at OWS & DDC controller				
Maintainability				
Accessibility				

Checklist for Pump and Pump Control

Items	Verification	Notes
Nameplate Data		
Pump Model #		
Serial #		
Class		
Type		
Manufacturer		
Motor volts/amps/kW		
Motor HP/rpm		
Model #		
Serial #		
Class		
Type		
Manufacturer		
Pump seal type		
Impeller diameter		
GPM		
Head-ft.		
Required NPSH		
Pump rpm		
Motor starter—amps		
Motor protection		
Constant/variable rpm		
Model #		
Serial #		
Class		
Type		
Manufacturer		
Variable speed drive		
Model #		
Serial #		
Class		
Type		
Options (PID and other)		
Manufacturer		

Test Data

Item	Design	Actual	Calibrated	Remarks
Measured Data				
Nominal data				
Speed (rpm)				
Volume (GPM)				
Discharge pressure				
Suction pressure				
Differential pressure				
Volts/frequency (V/Hz)				

Checklist for Pump and Pump Control—continued

Test Data—continued

Item	Design	Actual	Calibrated	Remarks
Measured Data				
KiloWatts/hour (kW/H)				
Phase 1,2,3 amps (A1/A2/A3)				
Reduced speed data at ....reduction:				
Speed (rpm)				
Volume (GPM)				
Discharge pressure				
Suction pressure				
Differential pressure				
Volts/frequency (V/Hz)				
Phase 1,2,3 amps (A1/A2/A3)				
KiloWatts/hour (kW/H)				
Reduced speed data at ....reduction:				
Speed (rpm)				
Volume (GPM)				
Discharge pressure				
Suction pressure				
Differential pressure				
Volts/frequency (V/Hz)				
Phase 1,2,3 amps (A1/A2/A3)				
KiloWatts/hour (kW/H)				
Reduced speed data at ....reduction:				
Speed (rpm)				
Volume (GPM)				
Discharge pressure				
Suction pressure				
Differential pressure				
Volts/frequency (V/Hz)				
Phase 1,2,3 amps (A1/A2/A3)				
KiloWatts/hour (kW/H)				
DDC Verification				
Name (acronym)				
DDC controller #				
Constant speed				
I/O module #				
Wire terminal #s				
Motor starter #				
Labeling				
Two speed				
I/O module #				
I/O module #				
Wire terminal #s				

Checklist for Pump and Pump Control—continued

Test Data—continued

Item	Design	Actual	Calibrated	Remarks
DDC Verification				
Wire terminal #s				
Motor starter #				
Labeling				
Status feedback				
Field device				
DDC controller #				
I/O module #				
Wire terminal #s				
Labeling				
Variable speed control				
DDC controller #				
Speed control I/O #				
Wire terminal #s				
Speed feedback I/O #				
Wire terminal #s				
VSC alarm I/O/ #				
Wire terminal #s				
Labeling				
Other I/Os				
Protocol interface				
Hardwired interlocks (name all)				
Labeling				
Application software				
Software interlocks (name all)				
Time schedule				
Other programs				
Prerequisites to S/S				
Alarm reporting				
Other associated items				
Witnessed at OWS & DDC controller				
Maintainability				
Access				
Maintenance instructions				
Lighting				

Engineering Tips Related to Pump and Pump Control Checklist

The most common pump types used in HVAC applications are:

1. Positive Displacement

   • Rotary

   • Piston

2. Centrifugal

   • Radial

   • Mixed flow

   • Axial flow

The most common operating speed of pumps is 1800 rpm but may differ from a low of 500 to a high of 3000 rpm.

The following are the most common pump equations:

$$\text{Capacity vs. speed: } \frac{\text{gpm}_2}{\text{gpm}_1} = \frac{\text{rpm}_2}{\text{rpm}_1}$$

$$\text{Capacity vs. impeller: } \frac{\text{gpm}_2}{\text{gpm}_1} = \frac{D_2}{D_1}$$

$$\text{Head vs. speed: } \frac{H_2}{H_1} = \left(\frac{\text{rpm}_2}{\text{rpm}_1} \right)^2$$

$$\text{Head vs. impeller: } \frac{H_2}{H_1} = \left(\frac{D_2}{D_1} \right)^2$$

$$\text{Break Horsepower vs. speed: } \frac{\text{bhp}_2}{\text{bhp}_1} = \left(\frac{\text{rpm}_2}{\text{rpm}_1} \right)^3$$

$$\text{Break Horsepower vs. impeller: } \frac{\text{bhp}_2}{\text{bhp}_1} = \left(\frac{D_2}{D_1} \right)^3$$

where

GPM = gallons per minute
rpm = revolutions per minute
D = impeller diameter (in.)
H = head (ft.w.g.)
BHP = brake horsepower

Head is a definition for different head pressures, such as friction head, suction head, dynamic discharge head.

Pump curves are normally furnished by pump manufacturers. They consist of groups of curves for different horsepowers, impeller diameters, and efficiencies. See Figure 8-21 for pump curves and an abbreviated curve for pump selection.

Figure 8-22 shows a design operating point on an abbreviated pump curve.

The following hydronic equations are most commonly used in system commissioning:

$$Q = 500 \times \text{GPM} \times \Delta t$$

$$\frac{\Delta P_2}{\Delta P_2} = \left(\frac{\text{GPM}_2}{\text{GPM}_1} \right)$$

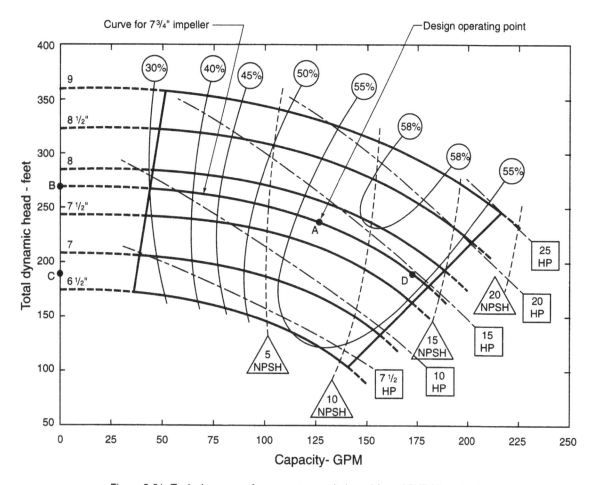

Figure 8-21. Typical pump performance curves (adapted from ASHRAE III, fig 8)

$$\Delta P = \left(\frac{GPM}{C_v} \right)$$

$$WHP = \frac{GPM \times H \times Sp.\ Gr.}{3960}$$

$$BHP = \frac{GPM \times H \times Sp.\ Gr.}{3960 \times E_p} = \frac{WHP}{E_p}$$

$$E_p = \frac{WHP \times 100}{BHP}\ \text{(in percent)}$$

where:

GPM = gallons per minute
Q = heat flow (btu/hr)
ΔP = pressure diff. (psi)
C_v = valve constant (dimensionless)
WHP = water horsepower
BHP = brake horsepower
H = head (ft w.g.)
Sp.Gr. = specific gravity (use 1.0 for water)
E_p = efficiency of pump.

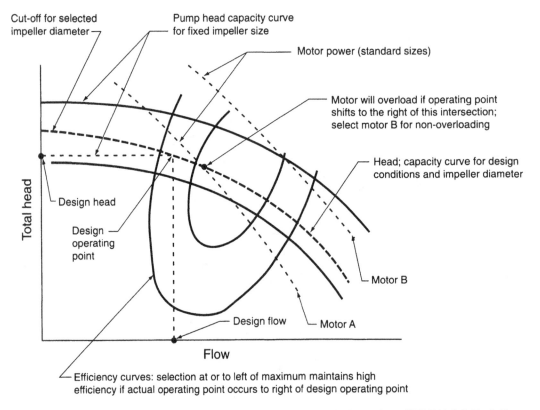

Figure 8-22. Design operating point on a pump abbreviated curve (adapted from SMACNA 8.8, fig. 8.6)

System curves, plotted against pump curves, show the effect of flow change on energy head pressure drop (see Figure 8-23).

$$\frac{H_2}{H_1} = \left(\frac{Q_2}{Q_1}\right)^2$$

Similar curves are used for open systems, such as cooling tower applications (see Figure 8-24).

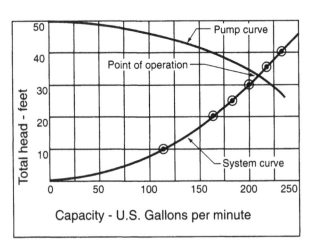

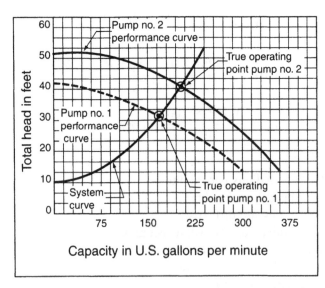

Figure 8-23. System curve plotted on a pump curve (adapted from SMACNA 8.8, fig. 8.7)

Figure 8-24. Open circuit system curve and true operating points for pumps 1 and 2 (adapted from SMACNA 8.10, fig 8.11)

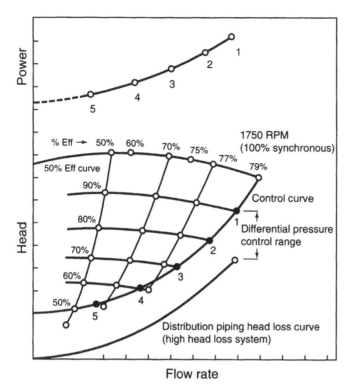

Figure 8-25. Pump control curve for variable speed drive (adapted from SMACNA 14.11, fig, 14-5)

Figure 8-25 is an example of Variable Speed Pump curves to obtain a "control curve."

Engineering Tips Related to Hydronic Systems Checklist

Hydronic systems utilized in HVAC applications are usually hot water and chill water systems. In some applications, glycol is being used in preheat coils and/or in heat recovery systems to prevent coil freeze up.

Hot water systems in commercial and industrial applications are pumped systems (forced). Most secondary systems are low temperature systems operating at a maximum temperature of up to 250°F and pressure up to 160 psi. Medium (up to 350°F/150 psi) and high temperature (350°F/300 psi) systems are utilized as primary systems. Hot water for HVAC applications can be generated by boilers on the premises or converted in a heat exchanger. The heating media in the hot water exchanger can be steam, hot water, or electricity. The most commonly designated hot water distribution systems are either one-pipe systems, two-pipe direct, or reverse return systems.

During commissioning, the commissioning agent should verify the function of all components of the system, including the hot water exchanger, piping systems, valves, strainers, expansion tanks, and positions of the balancing valves. The velocity limits should be between 1-4 feet per second (fps) with most systems designed at 2.5 fps for the upper floors. These velocities work also to reduce pipe noise caused by air and turbulence in the piping.

The head pressure drop at certain flow velocity can be determined from a catalog value for a certain coil by the following equation:

$$\frac{H_2}{H_1} = \left(\frac{Q_2}{Q_1}\right)^2$$

Checklist for Hot Water Zone Heating and Control

Items	Verification	Notes
Nameplate Data		
Hot water exchanger:		
Model #		
Serial #		
Type		
Size		
Manufacturer		
Rating (btu)		
Pressure/dp		
Capacity gpm		
Primary source		
Steam psi/lbsph		
Electrical V/kW		
Primary water temp/dT		

Test Data

Item	Design	Actual	Calibrated	Remarks
MEASURED DATA				
System pressure supply/return				
Temperature supply/return				
Flow rate supply/return				
Control set point				
Water make-up capacity				
Expansion tank capacity				
Variable speed pumping				
System differential pressure				
Balanced and marked:				
Radiator vents				
Balancing valves				
Circuit setters				
Draining valves				
Pressure reducing valves				
Strainers				
TAB reports verified				
DDC VERIFICATION				
Point inputs (acronym)				
DDC controller #				
I/O module #				
Wire terminal #s				
Labeling				
Point outputs (acronym)				
DDC controller #				
I/O module #				

Checklist for Hot Water Zone Heating and Control—contd.

Test Data—continued

Item	Design	Actual	Calibrated	Remarks
DDC VERIFICATION				
Wire terminal #s				
Labeling				
Application software				
Software interlock (name all)				
Time schedule				
Other programs				
Prerequisites to S/S				
Alarm reporting				
Other associated items				
Witnessed at OWS & DDC controller				
Maintainability				
Access				
Maintenance instructions				
Lighting				

where

Q = flow in (gpm)

H = head in ft w.g.

A pressure drop for a valve is calculated from a flow and Valve constant:

$$dp = \left(\frac{Q}{Cv}\right)^2 \text{ in (psi)}$$

where

Cv = Valve Constant

Q = flow in (gpm)

For better control, the steam hot water exchanger is controlled by 1/3 and 2/3 valves. The differential pressure between supply and return of some systems is maintained by differential pressure control valves; and some systems utilize variable speed drives on the pumps. Three-way valves can be used in the secondary distribution systems to maintain a constant hot water supply temperature in the system.

Checklist for Chill Water Systems and Control

Items	Verification	Notes
Nameplate Data		
Chiller/exchanger		
Pumps and motors		
Chillers, towers, DX units (use relevant checklists and manufacturer's documentation)		

Test Data

Item	Design	Actual	Calibrated	Remarks
Measured Data				
System pressure supply/return				
Temperature supply/return				
Flow rate supply/return				
Control set point				
Water make-up capacity				
Variable speed pumping				
System differential pressure				
Balanced and marked:				
Balancing valves				
Circuit setters				
Draining valves				
Pressure reducing valves				
Strainers				
TAB reports verified				
DDC Verification				
Point inputs (acronym)				
DDC controller #				
I/O modules #				
Wire terminal #s				
Labeling				
Point outputs (acronym)				
DDC controller #				
I/O module #				
Wire terminal #s				
Labeling				
Application software				
Software interlocks (name all)				
Time schedule				
Other programs				
Prerequisites to S/S				
Alarm reporting				

Checklist for Chill Water Systems and Control—continued

Test Data—continued

Item	Design	Actual	Calibrated	Remarks
DDC Verification				
Other associated items				
Witnessed at OWS & DDC controller				
Maintainability				
Access				
Maintenance instructions				
Lighting				

Checklist for Packaged Chillers and Control

Items	Verification	Notes
Nameplate Data		
Chiller make/model		
Type/size		
Capacity		
Refrigerant		
Starter V/A/kW		
Compressor make/model		
Type/size		
Suction temp/pressure		
Discharge temp/pressure		
Oil pressure		
V/A/kW		

Test Data

Item	Design	Actual	Calibrated	Remarks
MEASURED DATA				
Condenser				
Pressure/temperature				
Entering water temp/pressure				
Leaving water temp/pressure				
Differential pressure				
Differential temperature				
Evaporator				
Pressure/temp				
Entering water temp/pressure				
Leaving water temp/pressure				
Differential pressure				
Differential temp				
Flow				

Checklist for Packaged Chillers and Control—continued

Test Data—continued

Item	Design	Actual	Calibrated	Remarks
MEASURED DATA				
Compressor				
Low/high cut-out set point				
System pressure supply/return				
Temperature supply/return				
Flow rate supply/return				
Control set point				
System differential pressure				
Balanced and marked				
Balancing valves				
Circuit setters				
Draining valves				
Pressure reducing valves				
Strainers				
TAB reports verified				
DDC Verification				
Point inputs (acronym)				
DDC controller #				
I/O module #				
Wire terminal #s				
Labeling				
Point outputs (acronym)				
DDC controller #				
I/O module #				
Wire terminal #s				
Labeling				
Hardware interlocks				
Application software				
Software interlocks (name all)				
Time schedule				
Other programs				
Prerequisites to S/S				
Alarm reporting				
Other associated items				
Witnessed at OWS & DDC controller				
Maintainability				
Access				
Maintenance instructions				
Lighting				

Checklist for Cooling Tower/Condenser and Control

Items	Verification	Notes
Nameplate Data		
Make/model		
Type/size		
Capacity		
Refrigerant		
Water treatment		
Fan		
No. of fans		
Make/model		
Type/size		
Capacity		
Sheave diam/bore		
Motor		
Make/model		
Type/size		
HP/rpm		
V/A/kW		
Sheave diam/bore		
No. of belts/size		

Test Data

Item	Design	Actual	Calibrated	Remarks
Measured Data				
Water				
Entering water temp/pressure				
Leaving water temp/pressure				
Differential temp				
Differential pressure				
Flow rate				
Air				
Airflow rate				
Static pressure				
Average entering WB temp				
Average leaving WB temp				
Ambient WB temp				
Condenser				
Pressure/temperature				
Entering water temp/pressure				
Leaving water temp/pressure				
Differential pressure				
Differential temperature				
Balanced and Marked				
Balancing valves				
Circuit setters				
Draining valves				
Pressure reducing valves				

Checklist for Cooling Tower/Condenser and Control—continued

Item	Design	Actual	Calibrated	Remarks
Measured Data				
Balanced and Marked				
Strainers				
TAB reports verified				
DDC Verification				
Point inputs (acronym)				
DDC controller #				
I/O module #				
Wire terminals				
Labeling				
Point outputs (acronym)				
DDC controller #				
I/O module #				
Wire terminal #s				
Labeling				
Hardware interlocks				
Application software				
Software interlock (name all)				
Time schedule				
Other programs				
Prerequisites to S/S				
Alarm reporting				
Other associated items				
Witnessed at OWS & DDC controller				
Maintainability				
Access				
Maintenance instructions				
Lighting				

Checklist for Steam Heating Systems and Control

Items	Verification	Notes
Nameplate Data		
Pressure reducing station		
Valve make/model		
Type		
Entering psi/leaving psi		
Manufacturer		
Main Valve		
Make		
Type		
Size		
Manufacturer		
House Traps		
Make		
Model		
Type		
Manufacturer		
Condensate receiver		
Type		
Make		
Model		
Capacity		
Manufacturer		

Test Data

Item	Design	Actual	Calibrated	Remarks
Measured Data				
Calculated lb/hr				
Pressure min/max				
Flow min/max				
Radiator traps				
Riser traps				
Condensate receiver capacity				
Liquid mover capacity				
Balanced and Marked				
Radiator valves/traps				
Pressure reducing valves				
Strainers				
TAB reports verified				
DDC Verification				
Point inputs (acronym)				
DDC controller #				
I/O module #				
Wire terminal #s				

Checklist for Steam Heating Systems and Control—continued

Item	Design	Actual	Calibrated	Remarks
DDC Verification				
Labeling				
Point outputs (acronym)				
DDC controller #				
I/O module #				
Wire terminal #s				
Labeling				
Application software				
Software interlock (name all)				
Time schedule				
Other programs				
Prerequisites to S/S				
Alarm reporting				
Other associated items				
Witnessed at OWS & DDC controller				
Maintainability				
Access				
Maintenance instructions				
Lighting				

Engineering Tips Related to Steam Systems Checklist

Steam systems are utilized for district heating, as well as building heating and HVAC applications. Most steam distributions are at medium pressure (240-290°F/10-44 psi) or high pressure (300-355°F/55-125 psi). Low pressure steam (221-250°F/3-16 psi) is used for most HVAC applications.

Steam delivered to the coil or radiator at a certain pressure and temperature gives up its heat content, condenses, and returns as condensate to the boiler.

The amount of heat required to condense at 212°F/14.7 psi is 970.3 Btu/lb.

The condensate returned in [gpm] =
1000 [lb/hr]/8.33 [lb/gal] × 60 [min/hr]

The table in Exhibit 8-26 can be used as a reference for steam properties at different pressures.

Steam systems can be one-pipe or two-pipe systems. Condensate return systems can be gravity or vacuum systems. From most coils and radiators, the condensate is returned to condensate receivers or liquid movers, which then pump back the condensate via a distribution system to the boiler plant.

Pressure		Saturation Temperature °F	Specific Volume ft³/lb		Enthalpy Btu/lb		
Gauge psi	Absolute psia		Liquid V_f	Steam V_g	Liquid h_f	Evap. h_{fg}	Steam h_g
Vacuum 25 in. Hg	2.4	134	.0163	146.4	101	1018	1119
9.56 in. Hg	10	193	.0166	38.4	161	982	1143
0	14.7	212	.0167	26.8	180	970	1150
2	16.7	218	.0168	23.8	187	966	1153
5	19.7	227	.0168	20.4	195	961	1156
15	29.7	250	.0170	13.9	218	946	1164
50	64.7	298	.0174	6.7	267	912	1179
100	114.7	338	.0179	3.9	309	881	1190
150	164.7	366	.0182	2.8	339	857	1196
200	214.7	388	.0185	2.1	362	837	1179

Exhibit 8-26. Properties of saturated steam (reprinted from SMACNA pg. 16.35, table 16-36)

An important part of the steam condensate system is the steam traps, which keep the steam in the coil until it releases its latent heat, and release the condensate to the return system (see Figure 8-27).

Proper function of steam traps, strainers, and vacuum breakers should be checked, because they are the cause of most operating problems. Dirt in the pipes can clog up the strainer, or damage the trap. Leaky traps let live steam into the condensate system, causing uncontrolled heating of certain areas and possibly damaging condensate return pumps. Depending on the trap orifice diameter, the annual cost of steam losses can be quite substantial.

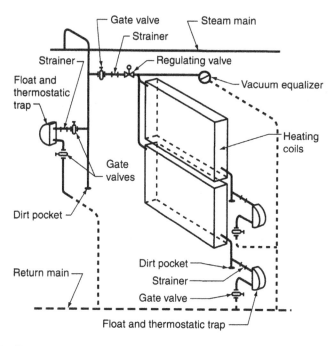

Figure 8-27. Typical connections of steam and condensate to a heating coil (adapted from SMACNA 9.17, fig. 9-17)

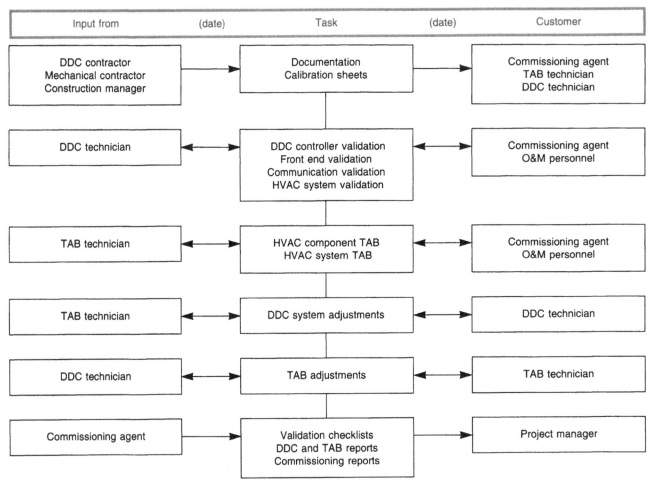

Input from	(date)	Task	(date)	Customer

| DDC contractor
Mechanical contractor
Construction manager | | Documentation
Calibration sheets | | Commissioning agent
TAB technician
DDC technician |

| DDC technician | | DDC controller validation
Front end validation
Communication validation
HVAC system validation | | Commissioning agent
O&M personnel |

| TAB technician | | HVAC component TAB
HVAC system TAB | | Commissioning agent
O&M personnel |

| TAB technician | | DDC system adjustments | | DDC technician |

| DDC technician | | TAB adjustments | | TAB technician |

| Commissioning agent | | Validation checklists
DDC and TAB reports
Commissioning reports | | Project manager |

Figure 8-28. Identifying customers

The proper removal of condensate (water) from coils is essential. Residual water in the coils can freeze in cold winter conditions (in steam pre-heat coils), or creates hammering noise any time steam hits the cold condensate water, especially during morning warm up. DDC control of steam (zone) valves should be slow at the beginning, to allow for warm up of the system, and then gradually faster to full opening. Leaky traps can be diagnosed by DDC systems utilizing temperature sensor(s) in the condensate return line(s) that analyze the condensate temperature rise over time.

Identification of Customers for the Commissioning Phase

The formal customers are identified by contractual relationships. The informal customers of the commissioning process are the owner's O&M personnel and future building occupants.

Commissioning Reports

The TAB reports for individual systems, distributions, and areas (rooms, zones, etc.) should be provided to the commissioning agent in a format suggested by National Environmental Balancing Bureau (NEBB). The DDC contractor should

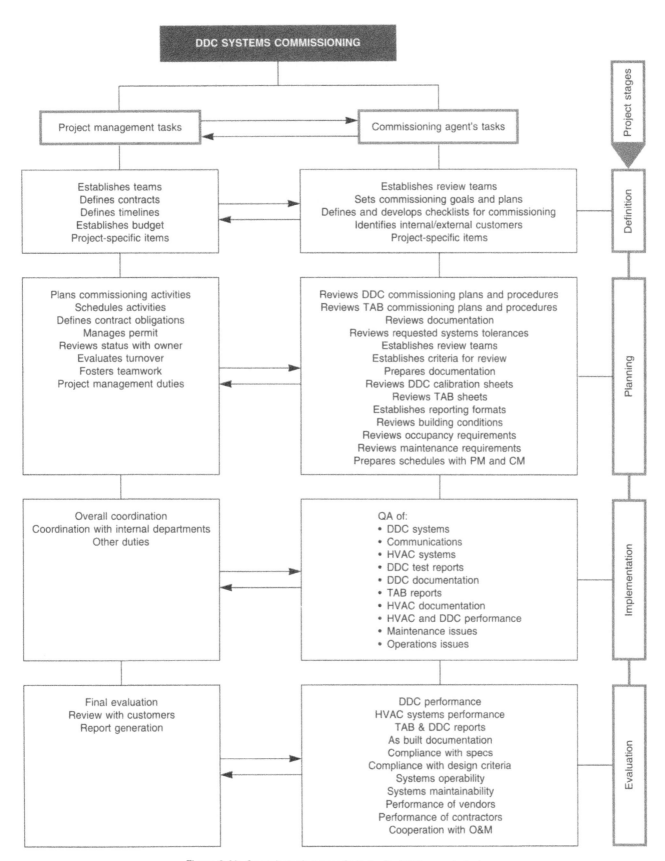

Figure 8-29. Commissioning agent's tasks for DDC commissioning

submit all calibration reports and software operating parameters in a format requested in the specification or by the commissioning agent.

The checklists provided in this chapter should provide the O&M personnel with unified commissioning documentation of the entire environmental control system. The data collected in the checklists provides facilities management with field verified data of the environmental control system and its components, and with initial start-up values. The collected data along with the systems documentation is utilized for operation and maintenance of the installed system, and as reference data for future re-commissioning.

Overview of Commissioning Agent's Tasks

The purpose of this overview is to provide the commissioning agent with a list of tasks at different stages of the commissioning phase of the project. In the definition and planning stages, the commissioning agent should customize the table for job- and site-specific conditions.

Chapter 9

Walk-Through and Systems Acceptance

Introduction

Environmental controls systems acceptance should be a continuous process throughout the design, development, and implementation phases of the project. The owner's representatives should be part of the entire process, participating in the review and approval process of every major step of the project. By the time the commissioning process is completed and the commissioning reports have been reviewed by the review team, there should be no surprises for the final walk-through. It is as if the job had been turned over to the owner's O&M personnel one segment at a time.

The final walk-through should be conducted jointly by the project manager, construction manager, and the owner's commissioning agent for the benefit of those who were not involved directly in the project review and commissioning process. Participation of O&M personnel who were not part of the global commissioning process should be encouraged. O&M personnel who have participated in the process are the best choice to present the operations and maintenance job details to their colleagues.

This approach is comforting to the future building occupants as well. Knowing that the O&M personnel have participated in the entire process and are familiar with the installed systems increases the occupant's confidence level, which is very important at the time of building turnover, when the expectations of the occupants are often conflicting with the initial "pains" of systems startups and component failures. Team approach and good communications with the building users may smooth out many potential conflicts.

Objectives for Final Walk-Through and Turnover

The main objectives are:

- Familiarization of O&M Personnel with the Completed Job
- Demonstration of the Job to the Building Users
- Formal Acceptance of the Job by the Owner

The chief objective is to demonstrate the completed and fully operational systems to the owner's internal organizations. If the Global Commissioning/TQM approach has been followed during the implementation process, the formal acceptance of the job by the owner becomes a formality. Such jobs should have a minimum or no turnover and acceptance problems. The final walk-through becomes a demonstration of a job well done, and the turnover becomes an administrative task conducted among the owner, construction managers, and contractors.

In a real life situation, the commissioning agent should expect comments and a "punch-list" from O&M personnel during the walk-through. A test operation of the systems during "real occupancy" can result in modifications and changes to the hardware or operating software prior to final acceptance. Such modifications should be completed as part of the performance contract, under extended warranty, or by other means, if the tasks are not included in the contracted scope of work.

Prior to final acceptance of the job, the project management should guarantee that the controls contractor will:

- Comply with the "punch list" and will complete all items by the set date
- Provide long-term performance testing and tuning of building environmental control systems to meet the design operating parameters for year round (four seasons) building operation.

A job walk-through will most likely result in comments and suggestions from the O&M personnel or building occupants. The "punch list" items should be corrected by the contractors as part of their contractual obligations. Items noted outside of the contracted scope of work should be evaluated by the project management, and if they are relevant to building operation (and if there is money left in the budget), they should be provided under separate change orders.

The most significant and most comforting arrangement for all involved parties is a contracted one-year systems performance warranty, as described in the DDC Performance Specification. Complex systems, such as environmental control systems, can be truly tested only in a real operating environment. The changing outside air conditions during four seasons in combination with real occupancy and changing heat loads of controlled spaces is a true test of the operating logic, equipment sizing, and systems reliability. Deviations from design, operating logic, set parameters, and other items may surface only under certain operating conditions. The good news is that most such "shortcomings" can be easily corrected by the participating HVAC, systems, and facilities engineers. Project management should anticipate this, and include the necessary time and money for performance warranty in the related contracts.

The walk-through and the final turnover can be considered complete upon acceptance of the following by the owner's project manager, commissioning agent, and O&M representatives

- HVAC systems demonstration, and final corrections of punch list items noted during the walk-through
- Controls and automation system operations and communications interface demonstrations, and final corrections of punch list items noted during the walk-through
- Turnover of HVAC and DDC as built documentation, as requested in the specifications and by the contract
- Turnover of relevant operating and maintenance manuals as requested in the specifications and by the contract
- Training of operators and maintenance personnel, as requested by the contract
- System acceptance by the owner's legal representation; start of warranty period and test operation

Performance Requirements and Measures

The commissioning agent should plan the final walk-through with all participants as described earlier. The following checklist should help the commissioning agent with a structured approach. Modify the checklist for site-specific conditions.

Checklist for DDC Walk-Through

Item	Approved	Not Complete	Completion Date	Comments
Field gear locations				
DDC cabinet locations				
Hardware accessibility				
Hardware serviceability				
Hardware labeling				
Wire labeling				
Color coding				
DDC task lighting				
AC power labeling				
AC breaker protective guards				
Hardware cut sheets				
Calibration sheets				
As built documentation				
Maintenance manuals				
Other				

Besides attending the formal walk-through, the project management and commissioning agent should also plan to meet with the building occupants. It is desirable to provide the building users with a description of the basic operating parameters (set temperatures, set relative humidities), operation modes (night setback, morning warm-up), how the occupants can reset their room temperature, override night setbacks from the local sensor, safety related issues (laboratory fume hood sash positions during unused and unoccupied conditions, purge in case of a toxic spill), and energy conservation issues (low air flow during night setback, use of light switch to activate occupied operating mode).

Some project managers underestimate the importance of developing positive attitudes in the building users toward the completed job and environmental systems performance. Moving into a new building or renovated space puts lots of pressure on the occupants of the building. Because of their own problems, uncertainties, and expectations, they can be agitated and sensitive to their new environment. Such pressures can contribute to a negative perception of the environmental control systems worsened by the usual aggravated start-up problems.

Project management can defuse potential problems, and build trust, by developing contacts with the department heads or other high level personnel, and by inviting the building occupants to an informal systems presentation. This should not be a selling job, but rather an account of features and capabilities of the installed systems. Building users appreciate the effort, and may contribute

241

ideas to improve the overall systems operation. User's acceptance of the installed systems is an important part of the acceptance process.

Test Operation, Extended Warranty

The long-term test operation of the installed environmental control systems is a transitional phase. The owner's O&M departments are responsible for the systems operation, and should be responding to the system problems, or calls from the occupants. They should also be monitoring the system's performance, and in cooperation with the vendor's systems engineer, should be adjusting the operating logic, set points, alarm limits, and other operations-related parameters to meet design criteria and occupancy comfort.

The DDC contractor's technicians should be called whenever there is a problem associated with components under regular warranty, as well as for problems associated with systems performance (if contracted under extended warranty), and systems engineering items. The long-range testing is to customize the systems to the operating conditions, and to expose systems or engineering problems during the warranty operation. It is important to focus on the system's performance in relation to meeting designed criteria.

Most "design and performance" problems should be eliminated during the extended warranty period. Otherwise, these problems could become chronic problems which cannot be corrected by regular systems maintenance. By the end of the warranty period, the operating logic should be perfected for all operating conditions, and the O&M personnel should have a good understanding of the systems operation, calibration requirements, shortcomings, frequent maintenance items, and other issues relevant to regular systems operation and maintenance.

The test operation and extended warranty is important for reducing future maintenance costs by focusing on regular operations and maintenance, rather than responding to problems which require engineering involvement and cannot be corrected by day-to-day maintenance.

During the test period, the O&M department should implement formal problem-reporting procedures. Every problem or suggestion should be recorded on Trouble Log Sheets, submitted to the engineering department for evaluation, and responded to by designated personnel. Each Trouble Log report should be followed up on, and corrective actions initiated. Properly documented problems will aid the maintenance department in analysis and classification of systems problems.

To make the process efficient, the designated engineers should always respond and follow up with appropriate actions:

- Approval of the corrective action taken
- Recommendation of a new corrective measure
- Changes in the operating software
- Follow up training for the operators or maintenance technician(s)
- Feedback to the DDC vendor
- Feedback to the person filing the Trouble Log Sheet
- Definition of a chronic problem for further management action
- Response to the occupants

Utilization of Trouble Log Sheets, over time, brings a systematic approach to maintenance procedures and practices. In most buildings and installed systems, problems tend to repeat themselves. Without a standardized approach to troubleshooting, a new method has to be invented every time there is a maintenance call. Facilities engineers should utilize Trouble Log Sheets in the development of standard approaches to documented problems.

The following is an example of a Trouble Log Sheet:

Environmental Controls Trouble/Enhancement Log Sheet

Fax to: **Fax #:**   **Phone #:**

Department: **Copy to:**

Building: **HVAC System:**

Area Served: **Date:**

DDC#:   **Application Specific Controller#:**

Problem description _____

_____

_____

_____

Corrective action taken _____

_____

_____

_____

Suggestions for improvement _____

_____

_____

_____

_____

Response

Reviewed by:

Response _____

Hardware modifications _____

_____

_____

Software modifications _____

_____

_____

Communications _____

_____

Change in the documentation _____

_____

Follow up action recommended _____

Environmental Controls Trouble/Enhancement Log Sheet—continued

Response

Response by the DDC vendor_____Under Warranty____P.O.#_____

Response by the mechanical contractor_____ HVAC under warranty_____

Specification of a failed component _____

_____P.O.# _____

Engineering redesign _____

By_____Time line_____P.O.# _____

Estimated cost _____

Funding source _____

Owner's Acceptance of the Job

Final acceptance should assure that all building systems meet designed performance criteria, and that together they provide a safe and comfortable environment for the building occupants, and an energy-efficient operation with minimum maintenance.

Depending on the nature of the job, there may be formal requirements for the issuance of operating permits, or safety certificates, or other requirements by outside or regulatory agencies. All certifications should be obtained, and permits issued, and turned over to the owner prior to formal acceptance of the job.

Identification of Customers

Formal customer relationships are established by the job contracts. Informal customers are the owner's O&M personnel, and the future building occupants.

The final turnover phase of the project has to satisfy both categories of customers.

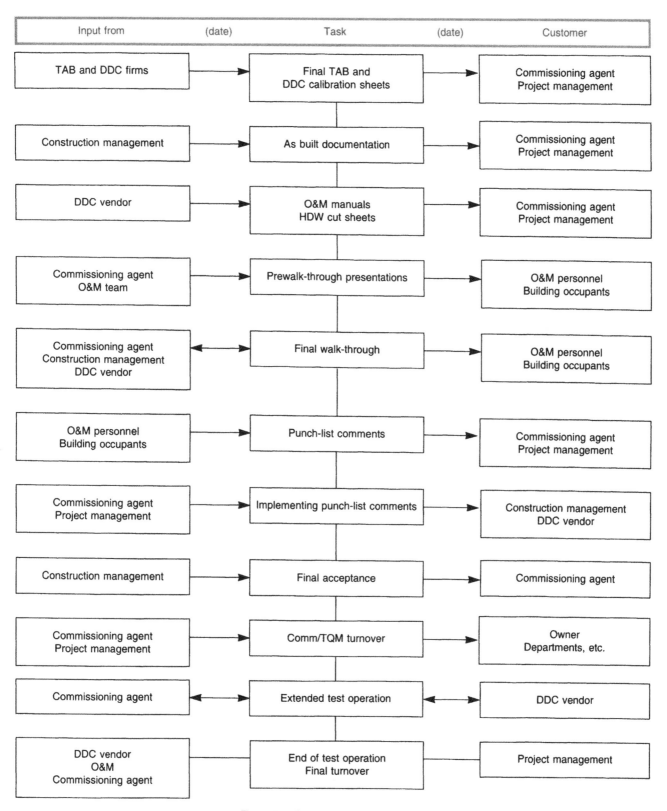

Input from	(date)	Task	(date)	Customer
TAB and DDC firms		Final TAB and DDC calibration sheets		Commissioning agent Project management
Construction management		As built documentation		Commissioning agent Project management
DDC vendor		O&M manuals HDW cut sheets		Commissioning agent Project management
Commissioning agent O&M team		Prewalk-through presentations		O&M personnel Building occupants
Commissioning agent Construction management DDC vendor		Final walk-through		O&M personnel Building occupants
O&M personnel Building occupants		Punch-list comments		Commissioning agent Project management
Commissioning agent Project management		Implementing punch-list comments		Construction management DDC vendor
Construction management		Final acceptance		Commissioning agent
Commissioning agent Project management		Comm/TQM turnover		Owner Departments, etc.
Commissioning agent		Extended test operation		DDC vendor
DDC vendor O&M Commissioning agent		End of test operation Final turnover		Project management

Figure 9-1. Customers at turnover phase

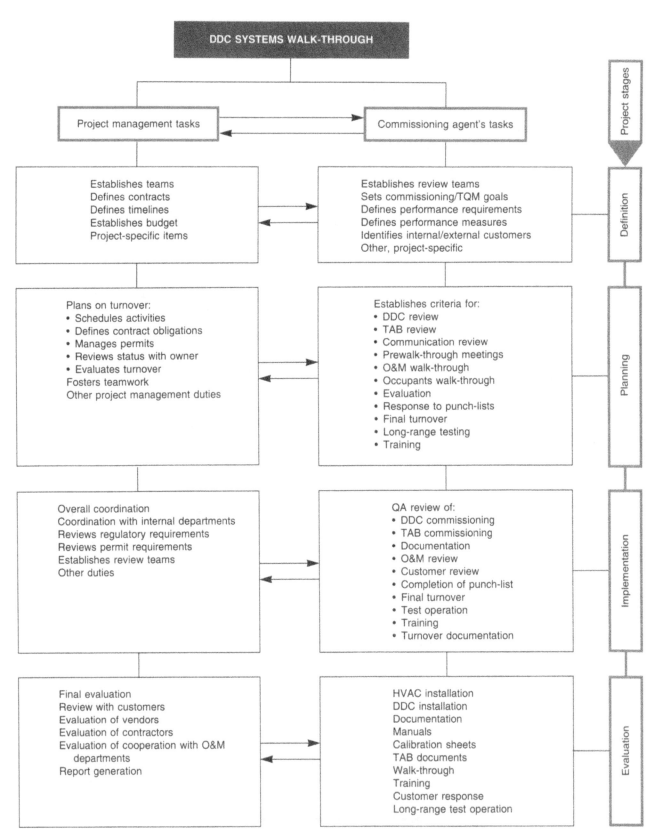

Figure 9-2. Commissioning agent's tasks for DDC turnover

Chapter 10

Training for Operations and Maintenance of Environmental Control Systems

Introduction

Environmental control systems are designed to provide and maintain safe and comfortable working conditions in buildings. Providing the design, implementation, and turnover all meet design criteria, the future performance of the building's environmental control system becomes the responsibility of the owner's O&M departments.

It should be emphasized that DDC systems are tools to be used to operate and troubleshoot building and environmental control systems-related problems. Therefore, the operators and maintenance training should not be restricted only to the DDC system, but should cover all associated HVAC systems. This also corresponds to the building occupants' perception of building environment problems. They are not concerned with whether the cold air coming out of the ceiling grills is a result of an undersized heating coil, inaccurate sensor, or faulty sequence of operation.

The training should cover all aspects of environmental systems (HVAC and DDC), their characteristics, operating logic, methods of troubleshooting, and service and repair of systems components.

DDC systems are the "brains" of building environmental control systems. As such, their software contains the definition of field points, and the operating logic for the connected mechanical systems. Environmental control system problems are either detected by DDC systems, or they are called in by building occupants in the form of requests for an increase or decrease in temperature, humidity, or air flow. Thus, building environmental problems are considered, and often treated, as if they are DDC problems.

In fact, most problems are related to the failure of HVAC hardware components. Some problems can be contributed to the application software. (However, if the software logic worked before, only operator entries or corruption of the software due to computer failure would cause software-related failures). A minimum number of problems are caused by DDC hardware or field gear failure. Reoccurring building problems should be considered chronic problems. They have their roots in building or systems design and implementation. Such problems can be eliminated only by re-design and modification of sections of the originally designed systems.

Prior to setting up a training program, it is important to acknowledge a few common characteristics that represent the most common pitfalls in the perception of environmental control systems. This knowledge may aid in understanding problems associated with environmental control systems, and reported by DDC systems:

1. Each DDC system on the market is unique. By implementing a system you are committed to the features and characteristics of that system. You can-

not expect an installed DDC system to have the best features of all DDC systems on the market.

2. DDC systems can control only what is connected to them. The architect or the owner often thinks of "fully controlled" buildings. However, the reality of budgetary constraints force engineers to "engineer out" some "nice to have" features and control points. In many instances, these points seem unimportant from the design standpoint, but are in fact very important from the operating and maintenance standpoints. Fortunately, all DDC systems have capacities to expand.

3. DDC systems cannot change the capacity and characteristics of connected mechanical systems. They can control them only at their maximum efficiency. A room served by a VAV box without a reheat coil will close the damper to restrict the air to the room, but cannot provide additional heating even though the DDC system reports a low temperature alarm from the room.

4. DDC systems cannot increase or decrease the air temperature, humidity, or air flow if there is not enough capacity in the controlled mechanical systems. The DDC system logic will control the AHU dampers and valves to provide adequate cooling. However, even with cooling coil valves commanded to 100% open, the room temperatures are going to rise quickly (on a hot summer day) if there isn't adequate cooling capacity in the HVAC system.

5. DDC systems control the connected mechanical systems as defined in the application software. Despite everyone's ideas on how the environmental system should operate, the DDC system will operate the connected HVAC systems according to the operating logic defined in its application software. The good news is that the operating logic can be modified at any time, on line, to provide the "most suitable operating logic."

6. Sequence of operation and the application software should be written jointly by the HVAC engineer, DDC systems engineer, and facilities engineer for optimum building operation. The combined experience of these engineers provides the most effective operating logic.

7. Most traditional controls problems are related to: (a) equipment failures, (b) operating logic, (c) building or systems design. Only a fraction of problems are DDC or "controls" hardware related.

8. DDC systems performance depends on its application engineering, and also on the skills of operating and maintenance personnel to operate the system effectively.

9. A well-engineered DDC system should be used as a tool to diagnose building and environmental control system problems. DDC systems can be programmed to control the connected HVAC systems, and also aid the operators and mechanics in HVAC systems diagnostics. How much online diagnostics is implemented depends on the level of expertise of the engineers involved in systems engineering.

10. Control technicians, usually, cannot fix building or systems design related problems no matter how often they are sent to the job site. Classifying these as "chronic" problems, re-engineering, and modifying such systems, can save a lot of headaches and payments for overtime.

DDC systems represent a major shift from hardware-oriented maintenance to on-line performance optimization, system diagnostics, problem analysis, and exchange of faulty hardware components. DDC operators and technicians should be selected and trained for operation and maintenance of environmental control systems. Their toolboxes should contain detailed systems knowledge, and they should have diagnostic tools loaded into their laptop PCs.

Selection of Candidates for the Training Program

There are two approaches in selecting candidates for DDC systems training programs:

a. Select candidates with electronics backgrounds and focus their training on HVAC applications
b. Select candidates with an HVAC background and focus their training on DDC systems and enhanced HVAC systems operation.

Both approaches can be equally successful if the candidates are properly cross trained. After completion of a comprehensive HVAC and DDC training program, both groups of candidates should be able to use the DDC system as a tool to analyze environmental control systems problems, optimize their operation, and minimize down time.

Prerequisites for Candidates

The following description could apply to environmental controls systems technicians as well as to DDC operators. Most of the real workload of DDC operators and technicians is associated with diagnostics of environmental control systems problems via the DDC system. Therefore, both groups, the operators and the service technicians, should be trained equally. However, facilities management should decide on the degree and intensity of training for the individual groups.

Prior Education

Service technicians or DDC operators should have a high school or trade school diploma, preferably in a related discipline. Additional training in HVAC or computer-related disciplines is advantageous.

Prior Experience

Two or more years of experience in servicing HVAC systems, refrigeration systems, or related control systems, preferably in a similar environment, is desirable. Additional experience with computers, chillers, boilers, and other major building equipment is advantageous.

Skills

Skills in analyzing problems in HVAC and other major building equipment are essential. Basic proficiency with computers and operating systems such as DOS is advantageous.

Service technicians should also be skilled in the diagnostics of field hardware using appropriate test instruments and using appropriate tools of the trade.

Benefits of Implementing a Training Program

1. Continuous education and enhancement of skills of the work force associated with the implementation of the new technology and the operation and maintenance of building environmental systems.

2. Transition from present hardware maintenance oriented approach to utilization of controls, monitoring, and diagnostic capabilities of DDC systems.

3. Transition from field to on-line calibration of connected analog points and on-line troubleshooting of connected mechanical systems.

4. Enhanced operation and reduction of customer complaints by on-line analysis and correction of problems before inconveniencing the building occupants.

5. Energy savings due to optimization of environmental systems operation and use of energy.

6. Reduction of maintenance costs due to online diagnosis of the faulty equipment, determination of its location, and determination of required tools and spare parts, prior to going to the job site.

Objectives for the Training Program

The training program should be based on the assumption that the candidates have the required education, experience, and skills.

1. Development of working knowledge for operation and maintenance of installed environmental control systems.

2. Adherence to procedures for responding to building environmental systems problems, as reported by DDC systems or by the customers.

3. Development of thorough knowledge related to operation, maintenance, and diagnostics of installed DDC systems.

4. Development of skills for on-line diagnostics of problems related to environmental systems operation.

Training Program Curriculum

The Structure

The proposed curriculum structure provides an overview of disciplines relevant to environmental control systems. It also reflects the shift from controls hardware maintenance and repair to on-line diagnostics of HVAC problems via the connected DDC system.

The training program can be divided into the following categories:

1. Buildings and design parameters
2. HVAC systems
3. DDC systems
4. Vendor systems specific training
5. O&M procedures
6. Building systems specific training
7. Pneumatic controls

The Curriculum

Buildings and Design Parameters (estimated minimum training time: 8 hours)

Building envelope (R-factors, U-factors)
Space utilization and exposure
Large building openings
Air intakes and exhausts
ASHRAE requirements for building design
BOCA and OSHA requirements for building design
Local code requirements for building design

Code requirements for heat loads, air changes, etc.
Interior zoning for HVAC systems
Site design parameters
Building pressurization requirements
Laboratory ventilation requirements

Building operation
Occupancy, standby, and night setback modes of operation
Summer/winter operation
Operations during occupied and unoccupied periods

HVAC Systems (estimated minimum training time: 32–40 hours)

Steam distribution systems
Pressure reducing stations
Steam coils and their controls
Steam traps and their maintenance
Vacuum systems and their control
Condensate return systems and their monitoring
Steam/condensate metering
Troubleshooting of steam/condensate systems

Steam generation
Boiler efficiency
Code requirements
Environmental requirements
Boiler control systems
Burner management systems
Water makeup and condensate return
Fuel selection and management
Energy management

Hot water distribution systems
Pressure/temperature reducing stations
System pressurization and balancing
Radiators/coils/FCUs and their control
Hot water exchanger and its control
Hot water pumping and its control
Hot water metering
Troubleshooting of hot water systems

Chill water distribution systems
System temperature, flow, pressure and balancing
Coils/FCUs and their control
Flat plate exchanger and its control
Chill water storage and control of charge/discharge cycle
Chill water pumping and its control
Chill water metering
Troubleshooting

Chiller types
Chill water and condenser water
Cooling towers and their control
Chill water primary and secondary pumping
Pump drives and their control
Efficiency calculations
Chiller plant optimization
Troubleshooting

Electrical systems
Wiring, installation, and labeling
Protective devices and their sizing
Switches, relays, contactors
Motors, drives, and their control
Transformers
Code requirements
Lighting requirements
Electrical metering
Troubleshooting of AC systems

Air Handling Units
Fans
Fan performance
Impellers, belts, and drive trains
Noise and vibration
Troubleshooting

Coils
Steam, HW, CHW coils
Coil characteristics
Coil stratification
Heat transfer
Coil strainers, drains, check valves
Troubleshooting

Air filters
Filter efficiency
Types of filters

Filter monitoring
Filter PM, cleaning, etc.

Dampers and grills, their types and functions
Air leakage and its effect on system efficiency
Preventive maintenance

Air distribution systems
Single duct systems and their control
Dual duct systems and their control
VAV systems and their control
Modification of constant volume to VAV systems
Mixing boxes
VAV boxes
Make-up air systems
General exhaust systems
Laboratory ventilation systems
Heat recovery systems

Minimum outside air requirements
Cooling with economizer cycle
Coil protection
Heating and ventilation systems

DDC systems (estimated minimum training time: 30–40 hours)

Field Controllers
Power supplies
Battery back-up
CPU
Communication modules
I/O modules
Signal conditioning
A/D converters
Wire termination
Grounding
Front end PCs
Printers

Communications
Communication modules
Communication modes (site & building level)
Communication speed at different levels
Communication error checks
Networks and network security

Communication protocols and their features
Drivers
Point mapping
DDC application software
Point definition
Alarm definition
Alarm message definition
Inhibit alarm functions
Hardwired interlocks
Software interlocks

Control end-to-end tolerance
PID loop calibration

Air control
Damper control
Mix air control
Economizer control
Preheat coil control
Cooling coil control
Reheat coil control
Fan control and feedback
Fan discharge temperature (FDT) control
FDT optimization using return air
PID loop control
Cascading individual loops
Calculations of set points
Online diagnostics and troubleshooting

Person Machine Interface (PMI) functions
System ID levels and access code
Operator Work Station (OWS) colorgraphics
Scanning points and their update on the CRT
Troubleshooting from the OWS
Trends and trend graphs
History data files
Report generation

Software updates, downloads/uploads, reboot, etc.
Database maintenance and update
Application software modifications

Call backs
Operators' trouble logs
DDC vendor support
Control Center documentation

Vendor Systems Specific Training (as offered by the vendors)

Can be accomplished on site, or at DDC vendor's office. DDC vendors regularly schedule training classes.

Operating and Maintenance Procedures (estimated minimum training time: 16 hours)

Procedures for:

- Operating system, monitoring, and alarm reporting
- Responding to heat calls
- Reviewing building systems operation
- Maintaining a trouble log
- Dispatching mechanics
- System troubleshooting from the OWS
- Generating reports (alarm report, point override, lockout, etc.)
- Maintaining application programs

- Updating database
- Maintaining documentation
- Handling heat calls
- Handling emergency calls
- Other site-specific items

Field procedures and areas of responsibility for:

- Trade assignments
- Problem diagnosis
- Repair of HVAC systems
- Troubleshooting DDC systems
- Troubleshooting the application programs
- Changing control parameters
- Tuning PID loops
- Field gear and end-to-end calibration
- Maintaining trouble logs
- Reporting systems modifications
- Reporting unsolved (chronic) problems
- Other site-specific items

Building Systems Specific Training

Architectural design, interior zoning, perimeter heating, HVAC systems design, DDC control systems, design parameters, etc., are all unique to the building. The best training is participation of O&M personnel in the design development process.

During Design

O&M personnel participate in review of:

- Design documentation
- Point list
- DDC design
- Operating logic
- PMI
- Software generation and testing

During Installation

O&M personnel participate in:

- Frequent visits to the job site to observe hardware locations
- Review of controls hardware installation
- Review of field gear installation
- Review of wiring, termination, and labeling
- Review of jumper settings, test voltages, calibration, etc.
- Review of software download to controllers
- Review of as built documentation

During Commissioning

O&M personnel participate in review of:

- Calibrations
- Tuning control loops
- Field hardware calibration
- Functional testing
- Test documentation

At System Start-up

O&M personnel participate in:

- Full participation during start-up
- Documentation review
- Walk-through
- Development of a punch list
- System acceptance

In Warranty Operation

O&M personnel participate with the controls contractor in problem solving and systems repair.

Pneumatic Controls (minimum training time: 16 hours)

- Review of pneumatic controls
- Sensors
- Actuators
- Controllers
- Air compressors
- Air dryers and filters
- Air distribution
- Vendor-specific issues

Training Sources

Professional Training Institutes

Professional training institutes are good providers of training. They specialize in, and focus on, particular segments of the industry, for example HVAC, pneumatic controls, boiler controls, or steam traps. The instructors are trained to present the material in a form understandable to the level of participants. Usually, instructors have a reasonably good understanding of the presented topics.

To plan an effective training program, the owner should discuss with the training institute modification of the topics to match the HVAC systems installed at the site. The owner's representative should participate in development of the curriculum, and demonstrate positive enforcement of the program. This method can be successfully applied for the HVAC and DDC curriculum listed on pages 251–254.

In-house Training

Large facilities have many experienced engineers and professionals capable (and willing) to put extra effort into the preparation and presentation of certain topics. The advantage of such training is in the intimate knowledge of the systems, and operating and maintenance practices, and in having a working relationship with the audience. Such training develops bridges between facilities engineers and operations, and brings the engineers closer to operations and maintenance personnel and their problems. These training sessions are always beneficial for both the instructors and the audiences.

The in-house training approach can be successfully applied for the building and design parameters, and O&M procedures sections of the program (items 1 and 5 in the previous section).

Utilization of Professional Training Books or Video Tapes

Training manuals and tapes, especially if they are issued by vendors as training tools, are very good sources of information, and affordable training tools. Other sources of published manuals and tapes are professional organizations such as ASHRAE, Instrument Society of America (ISA), and others.

This training material is very effective, especially if used in combination with an in-house "testing laboratory," where the technicians can try out the presented methods on a test bench. They are great for learning calibration procedures, the tuning of control loops, analog instrument validation, and so on.

This method can be successfully implemented for any of the categories or as an ongoing training method.

Vendor Training

Vendor training is an effective way of providing structured training on a particular system. This training has to be budgeted because it costs money for the classes plus travel expenses. Perhaps a more cost effective way is to have an on-site vendor training. DDC vendors provide such training for flat fees and a defined number of participants. The problem is in scheduling the participants so as not to impair normal operation.

This method can be successfully implemented for vendor-specific training categories (category 4 and 7 in the previous section).

On-the-Job Training

There is no better way to learn than by doing. Participating in a review of HVAC design and DDC application engineering serves two purposes: (a) assuring quality design which fits the O&M requirements; (b) learning the specific building system parameters, architecture, and operating logic at the early stage of the design/implementation process.

Participating in the installation and commissioning phase of the project, the O&M personnel learn a lot from observing the DDC field technicians at work, their methods of calibrations, tuning of control loops, and so on. The technicians also influence locating of hardware components and assuring adequate space and service access. They learn about the location of hardware components before they are being concealed by walls, ceilings, and other finishes.

This may sound like a waste of valuable maintenance time. In fact, this is the most cost-effective way of learning DDC systems applications, calibration, application software, and so on, as it relates to the installed systems. It makes the future service more effective because of the acquired knowledge, and the feel of ownership for the job.

Bibliography

1991 ASHRAE Handbook. Heating, Ventilating, and Air-Conditioning Applications. ASHRAE, 1991.

ANSI/ASHRAE 111–1988 Standard. Practices for Measurement, Testing, Adjusting, and Balancing of Building Heating, Ventilation, Air-Conditioning, and Refrigeration Systems. ASHRAE, 1988.

ANSI/ASHRAE 114–1986 Standard. Energy Management Control Systems Instrumentation. ASHRAE, 1987.

ASHRAE Guideline 1–1989. Guideline for Commissioning of HVAC Systems. ASHRAE, 1989.

ASHRAE Standard 135P, First public review draft. BacNet: A Data Communication Protocol for Building Automation and Control Networks. ASHRAE, Aug. 1991.

Barber-Colman. Controline (catalog), Barber-Colman Company, 1990.

Bernaden, A. Fohn, Williams, Anna Fay. Open Protocols: Communications Standard for Building Automation Systems. The Fairmont Press, 1989.

Boed, Viktor. "Controls and Automation Project Management." Yale Plant Engineering (internal publication), May, 1994.

Cilia, John. A guide for Building and Facility Automation Systems. The Fairmont Press, Inc., 1991.

Coad, J. William, PE. "Design/Build—An Option Not a Panacea." Heating/Piping/Air Conditioning, March 1994.

College of Engineering University of Wisconsin-Madison, Engineering Professional Development. Course material: Direct Digital Controls for HVAC Systems, 1993.

CSI. Performance Specifying Format. CSI, 1981.

Dunn, A. Wayne, PE, and Whittaker, John, PE, Ph.D. "Building Systems Commissioning and Total Quality Management-By Adopting TQM Principles, Engineers Can Design HVAC Systems That Provide Specified Performance at Affordable Cost." ASHRAE Journal, September, 1994.

Elovitz, M. Kenneth, PE. "Design for Commissioning, Successful HVAC System Designs Incorporate the Features Needed to Show Proper Performance During Commissioning and Daily Operation." ASHRAE Journal, October, 1994.

Geissler, Richard. "Reshaping the Way Controls Vendors Cooperate." Alliance, Spring, 1989.

Gustavson, Dale, CEM, CLEP. "Commissioning EMCS ... The Human Dimension," Strategic Planning for Energy and the Environment, Vol. 14, No. 2, 1994.

Hartman, B. Thomas. Direct Digital Controls for HVAC Systems. McGraw-Hill, Inc., 1993.

Honeywell. Serviceline Honeywell Direct Catalog, 11th edition. Honeywell, Inc., 1994.

Johnson Controls, Inc. Engineering Data Book. JCI, 1994.

_____. How to Choose the Right Facility Management System. JCI, 1993.

_____. HVAC/Refrigeration Controls Catalog. Johnson Controls, Inc., 1993.

Levermore, J. Geoffrey, Ph.D. "Commissioning Building Energy Management Systems—Commissioning Codes and Specifications for BEMS Are Needed so that HVAC Systems Will Better Meet the Needs of Building Occupants." ASHRAE Journal, September 1994.

NEBB. Procedural Standards for Building Systems Commissioning. NEBB, 1993.

Newman, Michael H. Direct Digital Control of Building Systems, Theory and Practice. John Wiley & Sons, Inc., 1994

Peters, Thomas J. Thriving on Chaos: Handbook for a Management Revolution. New York: Knopf, 1987.

_____, and Waterman, Jr., Robert H. In Search of Excellence: Lessons from America's Best-Run Companies. New York: Harper & Row, 1982.

Powers. Automatic Controls (catalog), MCC Powers, 1987.

SMACNA. HVAC Systems Testing, Adjusting & Balancing SMACNA, 1993.

Sterling, M. Elia, and Collett, W. Christopher. "The Building Commissioning/Quality Assurance Process in North America—The Commissioning Process Produces High Quality Buildings That Provide Comfortable and Healthy Conditions for Their Occupants." ASHRAE Journal, October, 1994.

Index

Printed and bound by CPI Group (UK) Ltd, Croydon, CR0 4YY

23/10/2024

01778246-0018